A QUANTUM WALK THROUGH SUSTAINABILITY

Nonlinear Thinking for a Nonlinear World

Paperback ISBN: 978-1-962081-46-7
Hardcover ISBN: 978-1-962081-51-1

The Press at Cal Poly Humboldt
Cal Poly Humboldt Library
1 Harpst Street
Arcata, California 95521-8299
press@humboldt.edu

Layout and Design by Mia Costales
Cover by Sanjay Wahal and Mia Costales

This work was developed with the assistance of artificial intelligence tools during the writing and editing processes.

A QUANTUM WALK THROUGH SUSTAINABILITY

Nonlinear Thinking for a Nonlinear World

Sanjay Wahal, PhD

The Press at Cal Poly Humboldt
Arcata, California

Praise for *A Quantum Walk Through Sustainability*

In *A Quantum Walk Through Sustainability*, Sanjay Wahal offers a bold and intellectually rigorous reframing of the sustainability challenge for an era defined by complexity and entanglement. By integrating systems science, climate dynamics, materials innovation, and digital governance, he moves beyond linear models and invites us to think in terms of adaptive, interconnected knowledge enterprises—an approach that resonates deeply with the institutional design principles we have advanced for transdisciplinary problem-solving at scale. Wahal demonstrates that effective climate action requires not only technological breakthroughs but new mental models capable of navigating feedback loops, tipping points, and emergent behavior. His synthesis of rigorous scientific and imaginative systems thinking provides a valuable bridge between research, policy, industry, and civic engagement. This book is essential reading for university leaders, policymakers, sustainability innovators, scientists, systems designers, and citizens committed to building resilient, regenerative futures in a nonlinear world.

—*Michael M. Crow, President, Arizona State University*

Dr. Sanjay Wahal has woven a fresh narrative of climate sustainability through the quantum metaphor. This perspective piqued my interest and kept me turning the pages with curiosity. Complex ideas, which are essentially nonlinear in nature, are presented with clarity, and the big issues are described with punch and conviction. As a fellow chemical engineer, I can clearly see how he has articulated global climate change as an amalgam of interconnected processes and systems. The practicality of the book makes it all the more credible. As the author writes, "It is an invitation to dance with paradox, to sit with ambiguity, and to act without full certainty." For those who are passionate about the direction of our planet Earth, this book offers a highly nutritious quantum of insight and learning.

—*Prakash R. Kota, Professor, University at Albany; MLPowersAI Founder*

With more than two decades of research in smart grids and electric vehicle integration, I find the "living grid," which frames distributed energy systems through the lens of ecosystem dynamics, a thoughtful and timely way to understand the evolving architecture of modern power systems. This book thoughtfully examines the role of computational intelligence in modern energy systems; from my perspective, advances in AI will further enhance the ability to coordinate and optimize distributed resources across future grids.

—*Rajit Gadh, Professor, UCLA; UCLA Smart Grid Energy Research Center Director*

A Quantum Walk Through Sustainability by Sanjay Wahal is a deeply thought out and groundbreaking exploration of the complex, interconnected challenges of the climate crisis and sustainability. With its creative use of quantum metaphors, the book offers a unique perspective on navigating nonlinear systems, emphasizing the importance of coherence, emergence, and adaptive thinking. Wahal masterfully integrates scientific rigor, systems theory, and actionable frameworks, making the book accessible to policymakers, educators, innovators, and citizens alike. It challenges traditional linear approaches to sustainability, inspiring readers to embrace uncertainty, foster resilience, and design for systemic transformation. This work is a clarion call to action, urging us to reimagine our relationship with the planet and become active participants in shaping a regenerative future. Truly a remarkable read.

—*K.J. Mulkherjee, Professor, Department of Biochemical Engineering and Biotechnology, Indian Institute of Technology Delhi*

In this interdisciplinary work, Dr. Sanjay Wahal offers a compelling and holistic perspective on sustainability and decarbonization. Rather than treating sustainability as a series of isolated steps, the book reframes it as a complex systems challenge—one that requires integrating engineering rigor, systems thinking, and human insight. It invites individuals, scientists, engineers, policymakers, and business leaders to approach the sustainability transition with deeper curiosity and more integrated, forward-looking solutions."

—Mahesh Gupta, Professor and Former Chair, Department of Mechanical Engineering, Kennesaw State University; President, Plastic Flow, LLC

Contents

Foreword for *A Quantum Walk Through Sustainability*

There are books that inform. There are books that illuminate. And every so often, there are books that expand the way we understand the world. *A Quantum Walk Through Sustainability* is one of those rare contributions. It brings together climate science, systems theory, quantum metaphors, and practical case studies into a unified, coherent narrative that feels both scientifically rigorous and urgently relevant.

Sanjay Wahal offers something unusual in the sustainability space. He offers clarity without oversimplification. He offers hope without naivete. He offers science grounded in data and models, alongside a framework that invites readers to think more expansively about the future. This balance is the strength of the work. It respects the complexity of the problems we face while equipping the reader with tools to navigate that complexity with understanding, imagination, and agency.

The central premise of the book is simple and profound. The climate crisis is not a linear challenge. It is a nonlinear, entangled, and adaptive system that behaves more like a network than a machine. It evolves through feedback loops, thresholds, delays, and emergent patterns. Traditional models that rely exclusively on direct cause and effect, straight lines, and predictable trajectories are no longer sufficient for the world we now inhabit. Sanjay makes this case with scientific precision and clarity. He invites readers to adopt a mindset informed by the dynamics of complex adaptive systems and enriched by the metaphors of quantum behavior, where superposition, entanglement, and probability offer parallel tools for thinking.

What becomes striking as the chapters unfold is Sanjay's ability to translate advanced concepts into language that scientists, policymakers, innovators, educators, and students can all understand.

He takes phenomena such as Arctic amplification, Amazon dieback, nonlinear energy transitions, and tipping cascades, and situates them within a broader conceptual frame that helps the reader see not just the parts, but the patterns. He explains why certain interventions create coherence and why others generate interference. He shows how decisions in one domain reverberate through others. He makes it clear that sustainability is not a technical problem alone. It is a systems problem, a relational problem, and a governance problem that spans biology, physics, economics, and culture.

The strength of this book lies in its ability to integrate. Sanjay does not ask the reader to choose between science and imagination or between quantitative models and human behavior. Instead, he demonstrates why all these dimensions must be understood together. Climate, energy, agriculture, industry, finance, and public policy are not separate chapters in the planetary story. They are interdependent systems that shape one another through both obvious and subtle channels. Sanjay provides the reader with a scientific vocabulary to understand these interactions and with a conceptual lens that brings coherence to the process.

One of the most valuable contributions of this book is the way it reframes the idea of solutions. In many public discussions, sustainability is presented as a set of discrete actions. Install solar panels. Electrify vehicles. Reduce emissions. Protect forests. Each of these is important. But Sanjay highlights something else that is equally important. The interactions among solutions often matter more than the solutions themselves. Technology deployed in isolation can fail when it is mismatched with surrounding systems. A policy implemented without understanding behavioral responses can generate unintended consequences. A material innovation that improves one part of the value chain can create new pressures elsewhere.

By contrast, solutions that are aligned with system dynamics can generate reinforcing feedback that accelerates progress. Sanjay describes these as quantum interventions, not because they rely on quantum mechanics, but because they operate with an understanding of entanglement, probability, and interconnected pathways.

This is where the book excels. It shows how to think, design, and act in a world that rarely behaves in straight lines.

Another strength of the work is the precision with which it addresses climate risk. Sanjay does not rely on vague warnings. He draws directly from climate physics, Earth system models, and decades of research on tipping points. The vignettes on the Arctic, the Amazon, and the Atlantic Meridional Overturning Circulation are grounded in scientific literature and presented with both technical accuracy and narrative clarity. These sections make the science accessible without diminishing its seriousness.

Yet even in these discussions of risk, the tone remains constructive. Sanjay emphasizes that nonlinear systems contain not only dangers but opportunities. Patterns can shift rapidly in positive directions when key leverage points are identified. Social norms can change suddenly. Technologies can diffuse faster than expected. Governance models can adapt. This perspective is crucial for readers who may feel overwhelmed by the magnitude of the climate crisis. The message is not that change is easy. The message is that change is nonlinear. Once systems begin to shift, they can accelerate in hopeful directions.

Throughout the book, Sanjay also pays close attention to the human dimension. He understands that data alone does not change systems. People change systems through decisions, values, culture, and collective action. The concept of the quantum citizen is one of the most compelling parts of the work. It reframes the individual not as a passive participant in large systems but as an entangled actor whose choices influence networks beyond their immediate perception. This framing honors both science and human dignity. It shows how agency operates within complex systems and how meaningful change often begins with informed, coordinated action at multiple scales.

The chapters on regenerative materials, adaptive energy systems, digital carbon tools, and polycentric governance demonstrate how science and design can converge to create practical paths forward. These are not distant or theoretical ideas. They are real developments already emerging across industries and communities. Sanjay outlines them with enough detail to inform experts while also making them accessible to readers who are new to the field. This balance is part of what makes the book so effective. It educates without overwhelming. It inspires without exaggeration. It grounds vision in evidence.

The final sections of the book are particularly powerful. They bring together the science of complexity, the lessons from quantum metaphors, and the reality of climate transitions into a coherent call to action. The message is that sustainability is not a linear endpoint to be reached. It is a pattern of adaptive relationships shaped by continuous learning. It is a form of choreography rather than control. It requires humility, foresight, and the ability to navigate uncertainty with intention. The reader is left with a sense that while the challenges are significant, the tools and insights needed to address them are already within reach.

What distinguishes this book is its unifying perspective. Many works on sustainability focus on technologies. Others focus on policy, finance, culture, or behavior. Few works bring all these together within a shared scientific framework. Sanjay does this with remarkable clarity. The result is a book that serves researchers, practitioners, educators, leaders, and citizens alike. It is a guide for those who want to understand the complexity of the planetary crisis and for those who want to contribute to solutions that match that complexity.

In a moment when the world urgently needs science that is clear, guidance that is coherent, and ideas that can bridge disciplines, this book stands out as both timely and essential. It helps the reader see the world as it really is. More importantly, it helps the reader imagine how the world might become. In the words of Peter Drucker, *the best way to predict the future is to create it*, a principle that sits at the heart of the approach Sanjay presents here.

Pete Dulcamara
Author, *High-Tech Heroes, Why Gen Z is our Last and Best Chance to Save the Planet*
Former Chief Scientist, Kimberly-Clark, Neenah, Wisconsin
2025

Prologue

A Walk Beyond the Obvious

We are living in a time that feels both electric with possibility and paralyzed by crisis. The twenty-first century has delivered us into a paradoxical era—a world where we are both more connected and more fragmented than ever, more capable and yet more vulnerable, more informed yet often misled by our own data and assumptions. We have scaled mountains of knowledge, built skyscrapers of innovation, and connected continents through invisible threads of digital information. And yet, in the face of the planetary emergency—climate breakdown, ecological collapse, and the gradual erosion of sustainability—we continue to remain linear in our thinking. We march forward with spreadsheets and strategies, as if the future were a straight line awaiting our arrival.

But the planetary future has not been unfolding as a straight line. It has twisted, folded, and jumped. It hides in the shadows of complexity and emerges through tipping points that defy our forecasts and control. What we are confronting is not simply a climate crisis, but a crisis of governance, coordination, and capacity to act on a scale. We are attempting to solve a nonlinear, entangled, and adaptive challenge with tools and mindsets that are still designed for compartmentalized, incremental responses. While certain elements of the climate crisis do require structured, linear solutions—like building renewable energy capacity, deploying infrastructure, or enforcing emission standards—these actions are only part of the picture. The greater challenge is that these 'linear moves' must be orchestrated within a dynamic, interconnected system, where solutions interact, amplify, or undermine one another. It is less like moving chess pieces on a grid and more like conducting a symphony where each instrument—policy, technology, finance, and social behavior—must stay in tune with the others while adapting to an ever-shifting rhythm.

This book begins with a simple yet provocative observation:

what if the crisis we face is less like a machine to be fixed and more like a quantum field to be navigated? A field where cause and effect are not neatly separated, where small signals can amplify into great shifts, where observation changes the reality being observed, and where possibilities exist in parallel until we choose which to bring into being.

A World Out of Balance

Consider the systems around us—our energy grids, food networks, urban ecosystems, financial markets, and cultural narratives. Each of these is interconnected with the others, and none operates in isolation. When a drought strikes, it is not just a water problem. It becomes a food crisis, a migration issue, a political flashpoint, and a financial shock. When a new technology like artificial intelligence emerges, it does not merely disrupt an industry; it reshapes labor markets, energy demand, cultural expectations, and even our notions of identity.

Our civilization, in its relentless pursuit of growth, has disrupted the delicate homeostasis that has sustained life on this planet for millennia. We have altered the chemistry of the atmosphere, the rhythms of the seasons, and the flow of rivers. We have turned forests into commodities, oceans into dumping grounds, land into landfills, and cities into heat traps. Yet, we still attempt to address these challenges through single-point interventions—planting a few trees, deploying a carbon tax, or building a new battery—believing that the sum of such isolated actions will help restore balance. However, this will not be adequate because each of these is *technically rational* within a narrow frame, but misses the entanglement of ecology, economy, and society. We too often act as if solving one node of the system will "restore balance," when in reality those interventions may shift, delay, or amplify other crises.

We need to move beyond singular solutions and linear fixes. What we require is not just a better set of tools, but an entirely new way of thinking—one that embraces the inherent complexity, interdependence, and dynamism of our world. Climate, ecology, economy, and society do not operate in isolation; they form an intricate web of feedback loops and cascading effects. Addressing this reality demands approaches that are adaptive, systemic, and capable of aligning multiple solutions that work together rather than in isolation.

The Quantum Lens

Quantum mechanics, one of the most profound scientific revolutions of the last century, revealed a reality far stranger and richer than the mechanistic universe of Newtonian physics. In the quantum world, particles exist in superposition, meaning they hold multiple states at once until observed. They are entangled, meaning the state of one affects the state of another, regardless of distance. And they are probabilistic, meaning certainty gives way to likelihoods and possibilities.

These principles are not just the domain of physicists—they are metaphors for how we might navigate the challenges of sustainability. **What if we learned to think in superpositions, holding multiple solutions open at once rather than collapsing too early into false binaries like "growth versus environment" or "technology versus nature"?** What if we understood our entanglement—not just with other humans, but with ecosystems, species, and the biosphere itself—as the foundation of collective action? And what if we accepted uncertainty, not as a weakness, but as the natural condition of a living, adaptive world?

A Personal Walk

The ideas in this book are not abstract theories for me—they emerge from decades of work at the frontlines of innovation, sustainability, and decarbonization. I have seen the power of cross-sectoral solutions, from bio-based materials that reimagine industrial supply chains to community-driven renewable energy systems that empower the marginalized. I have also witnessed how often brilliant technical solutions fail—not because they don't work, but because they are treated as isolated fixes rather than elements of a living, entangled system.

I wrote this book to offer a new vocabulary, a new map, and a new set of tools for those who care about building a sustainable future. Whether you are a policymaker, an entrepreneur, a student, or a citizen-activist, this book invites you to see the world differently and to act from that seeing.

A Quantum Walk Through Sustainability

This is not a book of answers, but of patterns. It is structured as a "walk," not a straight path. We will explore the inadequacy of linear thinking and the nature of complex adaptive systems[1]. We will dive into the metaphors of quantum mechanics—superposition, entanglement, the observer effect—and translate them into actionable frameworks for sustainability. We will examine case studies that reveal how systemic entanglement can both hinder and accelerate change. And we will end by imagining a future where leadership, citizenship, and technology operate not as separate levers, but as entangled forces that shape a coherent, regenerative world.

The chapters ahead are both reflective and practical. They challenge old assumptions while offering new tools for design, governance, and innovation. Throughout, we will return to a core insight: **sustainability is not a destination but a pattern of relationships—a dynamic, evolving dance between humans and the Earth.**

Why This Book, Why Now?

We are at a moment of deep inflection. The window for preventing catastrophic climate tipping points is closing, yet new opportunities for transformation are opening all around us—from digital infrastructure that can track carbon in real time to regenerative agriculture and circular materials that reimagine industrial metabolism. But these opportunities will only be realized if we can think and act beyond silos, beyond single solutions, and beyond the linear logic that got us here.

This book is my invitation to you—to walk with me through this quantum landscape. It is a journey through uncertainty, possibility, and systemic renewal. It is a call to embrace the paradoxes of our time, to see complexity not as chaos but as opportunity, and to become the coherence we wish to see in the world.

Shall we begin?

Preface

We are living in the Anthropocene—an age not defined by tectonic shifts or cosmic events, but by human action. Yet, it is not merely human influence that shapes our times; it is the convergence of crises—climate disruption, biodiversity collapse, resource depletion, the gradual loss of sustainability and institutional fatigue*. These forces do not unfold in neat, linear patterns. They cascade, compound, and collide in ways that defy prediction.

What we face is not a single crisis with a single solution. It is a **nonlinear crisis**—an entanglement of ecological, economic, and institutional tipping points that strain the adaptive capacity of our linear systems.

We define this nonlinear crisis as:

"The loss of planetary homeostasis caused by entangled ecological, climatic, and institutional tipping points—exceeding the adaptive capacity of linear systems."

This is not a gradual slope of unsustainability but a cliff of compounding breakdowns. Our models, our governance frameworks, and even our decision-making tools struggle to keep pace with the volatility of this moment.

Yet our approaches to sustainability often rely too heavily on yesterday's logic. We lean on spreadsheets to model dynamic realities like wildfires, or on pilot programs to test planetary-scale interventions. While these tools are necessary and valuable for certain aspects of problem-solving, they are often inadequate on their own to

* **Institutional fatigue** refers to the declining capacity, effectiveness, and legitimacy of the political, economic, and social institutions we rely on to manage collective challenges. It is not just a matter of bureaucratic inefficiency or leadership failure; it is the systemic exhaustion of governance structures, global organizations, and regulatory mechanisms under the weight of complex, interlinked crises. When institutions are strained by multiple crises at once, they become reactive rather than proactive, slow to adapt, and unable to foster the long-term vision needed to address global challenges such as climate change.

capture the complexity, feedback loops, and systemic risks inherent in the climate crisis. This book was born out of a recognition of these limitations and a deep belief that we must complement these traditional tools with new frameworks and ways of thinking—ones capable of addressing the nonlinear and interconnected nature of sustainability challenges.

From Molecules to Systems

Over the past three decades, I have had the privilege of working across the arc of innovation—from high-performance materials to bio-based polymers, from multinational corporations to start-ups, from cleantech strategy to advisory. My work has spanned molecules to markets, theory to technology transfer, and systems modeling to venture creation.

Most recently, my focus has been on enabling structural decarbonization—critically reviewing breakthroughs such as gene-edited low-lignin trees[2] and advancing systemic interventions like the eCarbon Card[3] for India, designed to bring carbon accountability to the individual level through digital tools. Across these diverse efforts, a single truth has become clear: **linear thinking is inadequate when confronting the realities of complex, interconnected systems.**

Why Linear Thinking Fails

Climate change is not a single-variable challenge to be solved by linear cause-and-effect logic[4]. It is a **complex adaptive system**[5]—riddled with feedback loops, thresholds, and uncertainties. Its behavior emerges from an intricate web of interdependent factors:

- **Anthropogenic Drivers** – Emissions (CO_2, CH_4, N_2O), deforestation, industrial activity.
- **Natural System Dynamics** – Ocean currents, ice melt, soil-carbon cycles.
- **Feedback Loops** – Permafrost methane release, sea ice loss reducing albedo.

- **Socioeconomic Forces** – Global trade, urbanization, energy inequities.
- **Institutional Constraints** – Policy fragmentation, governance delays.
- **Cultural Dynamics** – Misinformation, behavioral inertia, polarized perceptions.

These variables do not act in isolation. They coalesce into emergent patterns—sometimes stabilizing, sometimes amplifying—making linear predictions dangerously insufficient. Solving climate change requires **whole-system design, plural knowledge frameworks, and adaptive governance.**

In Crisis, Opportunity

But within this turbulence lies possibility. We are not merely solving for emissions; we are redefining how humanity lives within planetary boundaries. This demands that we unlearn outdated mental models. It calls for the courage to embrace uncertainty, paradox, and emergence rather than forcing the world into linear molds.

This is where **quantum thinking** enters—not in the narrow, physical sense, but as a metaphor for interconnectedness, superposition, and probability. Quantum thinking allows us to navigate solutions that are simultaneously local and global, economic and ecological, technological and cultural. It invites us to move from control to participation, from predictability to adaptability.

A Quantum Walk

This book is not a manifesto or a blueprint. It is a **walk—a quantum walk—through paradoxes, solution spaces, and reframed questions.** It is an invitation to see yourself not just as a consumer or policymaker, but as a **quantum actor in an entangled ecological and economic system.**

My hope is that this book sparks conversations that transcend the binary, inspires systems-level action, and helps you see that solving our planetary crisis will require both a telescope and a microscope—and a prism that refracts light in unexpected directions.

This journey is not toward definitive answers but toward deeper possibilities.

This is a quantum walk through sustainability.

Vignette Prologue – Quantum Snapshots of a Nonlinear World

The world we inhabit is not a machine of predictable gears and levers—it is a living, breathing network of interdependent systems. Climate, oceans, forests, and societies are not separate chapters but overlapping verses in the same song. Each pulse of heat, each gust of wind, each policy delay sends ripples through this network, shaping probabilities rather than certainties. This is why the crises we face today cannot be understood as isolated events. They are **quantum events**—nonlinear, entangled, and capable of cascading into outcomes we can scarcely imagine.

To step into this quantum lens is to see that the Earth's stability is maintained not by a single pillar but by a web of feedback loops—loops that can stabilize but also amplify. When one loop falters, it can send shockwaves across continents and decades. A glacier melting in the Arctic may alter a farmer's harvest in India; a fire in the Amazon may shift rainfall in West Africa. These are not distant or disconnected phenomena. **They are the probabilistic ripples of an entangled planet.**

The following vignettes are not just scientific snapshots; they are stories of coherence and collapse, of how living systems respond when stressed beyond their thresholds. Think of them as quantum snapshots of a nonlinear world—a way to feel, not just analyze, the fragile balance we are tipping. Each vignette is a window into a different planetary organ: the Arctic as the reflective mirror, the Amazon as the lungs, and the Atlantic currents as the heartbeat. Together, they remind us that to navigate this future, we must think less like engineers of control and more like participants in a living quantum dance.

Vignette 1: The Vanishing Mirror – Arctic Feedback in Motion

The Arctic, once Earth's gleaming mirror, is losing its reflective edge. For thousands of years, its ice sheets served as a planetary albedo shield—casting sunlight back into space and regulating Earth's thermal balance. But now, this mirror is cracking, and with it, our illusion of climatic stability.

Each summer, more ice vanishes than returns. Where white ice once reflected heat, dark ocean waters now absorb it, amplifying warming and accelerating ice loss—a feedback loop that strengthens itself with each passing year. The Arctic is not melting linearly; it is quantum-leaping toward a new state, where small increments of change unlock disproportionately large consequences.

The thawing permafrost, a frozen archive of ancient carbon, is releasing methane—an invisible accelerant that is 80 times more potent than CO_2 over a 20-year horizon. Dormant microbes awaken and metabolize organic matter, emitting greenhouse gases in a self-perpetuating spiral. Hillsides in Siberia collapse into thermokarst craters, creating "drunken forests" that lean like collapsing dominoes.

From a systems perspective, the Arctic is not a remote casualty; it is an active agent of global destabilization. Its shifts distort jet streams, sending polar air deep into temperate zones, triggering floods, cold snaps, and heatwaves in places unprepared for such extremes. Indigenous communities lose ancestral hunting grounds, while wildfires sweep across tundra landscapes, turning carbon sinks into carbon sources.

In quantum terms, the Arctic is entangled with every corner of the planet. Its transformation is not a single cause with a single effect but a cascade of probabilities.

Reflection Prompt

If the Arctic is the Earth's thermostat, how do we navigate a world where the "temperature dial" is no longer linear but chaotic and entangled?

Vignette 2: The Dying Lungs – Amazon Dieback in Real Time

For centuries, the Amazon rainforest has been the planet's largest living engine of renewal—a breathing system of 390 billion trees that inhale carbon, exhale oxygen, and regulate rainfall like a global set of lungs. Today, those lungs are beginning to wheeze.

Deforestation—driven by roads, ranching, logging, and monoculture—has stripped away over 17% of the forest in just 50 years. Fires, some deliberate and others born from drought-stressed debris, scar the canopy. But the deeper crisis lies in the **feedback of broken cycles.** The Amazon recycles its own rain through evapotranspiration: trees release moisture that forms clouds and feeds rain back into the forest. Remove too many trees, and the feedback loop collapses. Large swathes of the Amazon are already transitioning into **savannas**, losing their carbon-storing capacity and resilience.

Satellite data now show eastern regions of the Amazon acting as net carbon emitters. Scientists warn that if deforestation crosses 20–25% of total forest cover, a **tipping point**[6, 7, 8, 9, 10, 11] could trigger irreversible dieback—a rapid flip from rainforest to dry scrubland. This is not a slow death; it is a **quantum collapse** of a complex system.

The Amazon's fate is not confined to South America. Its rainfall patterns influence agriculture in North America and Africa. Its role in stabilizing the global carbon and water cycles is unmatched. Should it fall, up to 90 billion tons of CO_2 could be released—equivalent to more than a decade's worth of global fossil fuel emissions.

In quantum terms, the Amazon is a node of planetary entanglement. When it "stops breathing," the shock ripples through food systems, weather patterns, and economic networks worldwide.

Reflection Prompt

What happens when the world's most complex ecosystem—a living quantum network—loses coherence and stops breathing?

Vignette 3: The Slowing Pulse – Atlantic Currents in Distress

Beneath the surface of the Atlantic lies an unseen yet colossal conveyor belt—the Atlantic Meridional Overturning Circulation

(AMOC). For millennia, this circulation of warm and cold water has been the **heartbeat of the planet's climate system**, regulating heat distribution, nutrient flows, and oceanic rhythms across hemispheres.

But the pulse is weakening.

The melting of Greenland's ice sheets floods the North Atlantic with freshwater, diluting its salinity and weakening its density. This prevents the cold, salty waters from sinking as they once did, slowing the currents that carry warm tropical water northward. This is not a gradual slowdown—it is a nonlinear weakening with the potential for sudden collapse.

Paleoclimate records tell us that when AMOC has faltered in the past, Europe plunged into ice-age conditions, African monsoons shifted, and global weather patterns convulsed. Current observations suggest AMOC is at its weakest in over a thousand years. Some models warn of a collapse within this century if greenhouse gas emissions continue unchecked.

The consequences of an AMOC collapse are not local but **planetary in scope.** Northern Europe could cool even as the rest of the planet warms. Tropical rain belts could migrate, inducing severe droughts in the Amazon and West Africa. Fisheries across the Atlantic could collapse due to disrupted nutrient flows. Sea levels along the U.S. East Coast could rise faster than global averages.

Like a heart with an irregular rhythm, the AMOC gives subtle warnings—until the arrhythmia becomes catastrophic.

Reflection Prompt

What does resilience look like when the planet's thermal heartbeat falters—and can we learn to adapt to its quantum rhythms?

Closing Paragraph: These vignettes are not isolated warnings but signals of a deeper truth: we are living inside a web of entangled crises where small perturbations can ripple across continents and decades. To navigate this reality, we must shed the illusion that sustainability can be engineered through linear fixes or siloed policies. Instead, we need a mindset attuned to complexity, emergence, and possibility—a quantum mindset. This book invites you to walk through that lens, to see how the smallest shifts in design, culture, and governance can reverberate like quantum particles, amplifying change across the system. What follows is not a search for definitive answers but a journey through patterns, paradoxes, and pathways—a quantum walk through sustainability.

Introduction

Navigating Complexity with Nonlinear Imagination

We are no longer dealing with a climate crisis that can be solved by linear equations or technological checklists only. The sustainability challenge we face today is not just a matter of emissions reduction or policy enforcement—it is a full-spectrum disruption of ecological, economic, and institutional systems. Solving it requires a different kind of logic. A different way of seeing. A different way of walking.

This book proposes that way: a quantum walk through sustainability.

Rather than offering prescriptive solutions, this book invites readers into a new mental model—one that draws from systems thinking, quantum metaphors, and complexity science to understand and intervene in the messiness of the real world. To win this battle, we need nonlinear thinking[12]. In contrast to traditional climate literature that treats sustainability as a linear problem of inputs and outputs, we explore it here as an entangled, adaptive, and emergent journey.

Much like the quantum walk in physics, where particles explore multiple paths simultaneously, our sustainability solutions must also be multidimensional—embracing uncertainty, contradiction, feedback, and co-evolution.

Historical Note: Humboldt's Quantum Vision of Nature

It is tempting to think of "quantum" or systemic thinking as a modern innovation born from complexity science and nonlinear dynamics. Yet the idea that nature operates as an interconnected web has deep roots. Over two centuries ago, **Alexander von Humboldt**

(1769–1859)[13]—often called the father of ecology—described the Earth as a living whole where climate, vegetation, soils, and human activity interacted through feedbacks and dependencies.

Humboldt's writings emphasized that interventions in one domain—such as deforestation in the Andes—could cascade through others, altering water cycles, soils, and even regional climate[14, 15]. In doing so, he anticipated what today we would call **entanglement** and **systemic causality**. His vision stood in contrast to the linear, mechanistic worldview that rose with the Industrial Revolution and dominated much of 20th-century policy and economics.

By revisiting Humboldt, we see that systemic thinking is not new—it is a rediscovery. Modern tools of AI, IoT, and nonlinear modeling simply extend what Humboldt grasped intuitively: that Earth is a complex, dynamic system where small causes can trigger disproportionate effects. In many ways, the "quantum sustainability" lens advanced in this book is less a rupture with history than a return to its deepest insights.

This book is a journey across the tangled terrain of sustainability—not as a static checklist of climate solutions, but as an evolving exploration of systems, paradoxes, and emergent possibilities. This is not a manual of control. It is a guide to choreography. It aims to equip readers with the intellectual tools, design frameworks, and mental flexibility to respond to the climate crisis not with rigidity, but with resonance. The chapters that follow are designed not to deliver fixed answers, but to provoke nonlinear questions, offer new vantage points, and support more adaptive responses to a rapidly changing world. Here is a brief preview of the path ahead:

Structure and Purpose of This Book

The book unfolds in five parts, walking the reader through a reframe of the crisis, an introduction of quantum thinking, proposing solutions beyond the obvious as a reimagination of the tools, and a reconstitution of governance / agency and finally proposing design for the quantum leap. Next is a brief preview of the journey:

Chapter 1: From Random Walks to Quantum Leaps

We begin by tracing the intellectual origins of this work—from the financial metaphor of the "random walk" to the quantum logic of Schrödinger and the systems wisdom of Donella Meadows. This chapter explains why sustainability can no longer rely on linear, deterministic models only.

Chapter 2: The Failure of Linearity

Linear tools and projections may fail to anticipate runaway feedback loops, social tipping points, and delays in response. This chapter examines the limitations of traditional thinking and why a new paradigm is needed.

Chapter 3: Climate Crisis as a Complex Adaptive System

Here, we introduce *complexity science.* We explore the climate not as a closed problem, but as a dynamic, adaptive, and nonlinear system—with Arctic amplification and Amazon dieback as instructive case studies.

Chapter 4: What Is a Quantum Walk?

This central metaphor offers a new language for sustainability. Drawing from principles like superposition, interference, and entanglement, we reinterpret sustainability as a multidimensional, probabilistic challenge.

Chapter 5: From Leverage Points[16] to Entanglement

Revisiting Meadows' leverage points, we show how deep change requires paradigm shifts and the reframing of relationships, not just metrics. Entanglement or targeted entangled intervention—entangled leverage, rather than isolated intervention, becomes the design principle.

Chapter 6: Systemic Entanglement

Using the lens of food, energy, water, carbon, and justice, this chapter explores how systems are deeply interwoven—and why isolated climate solutions (like EVs or solar panels) can create new unintended consequences.

Chapter 7: From Linear Solutions to Quantum Interventions

Through industrial case studies—like gene-edited low-lignin trees[17] and seaweed-based fibers[18]—we demonstrate how nonlinear material innovations are opening up regenerative pathways.

Chapter 8: Digital Permutations

We explore the digital dimension of systems change: how blockchain, behavior-based nudges, and carbon tracking tools can coordinate decentralized, real-time feedback loops for collective action.

Chapter 9: Bio-Based Futures and Material Metamorphosis

This chapter maps the rise of regenerative materials—from mycelium composites[19, 20] to agro-waste foams[21] to coffee waste-constituted concrete[22, 23] to carbon-sequestering algae[24, 25, 26, 27, 28, 29, 30]—offering a vision of material economies aligned with circularity and ecosystem logic.

Chapter 10: Energy Systems as Living Systems

We reimagine energy infrastructure as an adaptive, modular ecosystem—coordinated not centrally, but through AI, sensors, and community-owned microgrids.

Chapter 11: Finance as a Field Function

Finance is not just a neutral enabler—it shapes time, risk, and decision horizons. We explore tools like carbon indices, climate credit cards, and swarm capital to align finance with nonlinear reality.

Chapter 12: Quantum Governance and Polycentricity

We expand on governance models that mirror ecosystem logic—distributed, decentralized, and self-organizing—allowing coherence without central control.

Chapter 13: Carbon Indices and Market Intelligence

We examine how carbon indices can act as living signals—integrating prices, policy momentum, technological change, and narrative sentiment into a multidimensional compass for climate action. By pairing market intelligence with real-time feedback loops, these tools can guide adaptive strategies, democratize access to carbon data, and align global actors toward a just, coordinated transition.

Chapter 14: Time, Uncertainty, and Climate Leadership

Leadership in a nonlinear world requires humility, foresight, and comfort with ambiguity. We explore anticipatory governance, scenario constellations, and the mindset of nonlinear leaders.

Chapter 15: Designing for Emergence

Rather than planning with fixed endpoints, we discuss how to design ecosystems that evolve over time—via feedback loops, prototype portfolios, and learning architectures.

Chapter 16: The Role of the Quantum Citizen

Systems do not change without shifts in identity. This chapter centers the individual as a node of systemic influence—capable of sense-making, narrative-shifting, and regenerative agency.

Chapter 17: Walking the Future Backward

We conclude with vision-driven planning: how to backcast from preferred futures and create coherence across the quantum possibilities of today.

This Book Is a Prism, Not a Prescription

Quantum Walk Through Sustainability is not a linear read—it is an invitation to dance with paradox, to sit with ambiguity, and to act without full certainty. It speaks to policymakers, innovators, designers, and everyday citizens seeking to make sense of a world that defies easy answers.

It is not a toolkit. It is a telescope.

Not a formula. But a forest.

And above all, not a fixed doctrine—but a dynamic orientation toward co-creating a livable, regenerative, and just future on a shared, beautiful planet.

Conclusion: Embracing the Nonlinear Future

We close with a synthesis: not a call to control complexity, but to choreograph it with humility. The path forward demands nonlinear thinking, quantum awareness, and systems imagination.

This is not a manual. It is a mirror. A microscope. A prism. An invitation to reframe how we live, lead, and act in an entangled world.

Author's Note

Why I Wrote This Book—and for Whom

This book was born not from certainty, but from disquiet.

As someone who has spent decades at the intersection of science, industry, systems thinking, and innovation, I've seen firsthand the brilliance—and blindness—of our dominant approaches to sustainability. I've watched policy cycles repeat, pilot programs plateau, and bold visions get entangled in bureaucratic or market inertia.

But I've also witnessed moments of catalytic change—when a single innovation, a single policy, a single community decision ripples across networks and unlocks new behavior at scale. I've seen the power of reframing—from control to emergence, from linearity to complexity, from isolation to entanglement.

This book is an offering to that *possibility space.*

It is a product of my lived experiences:

- as a technologist who has seen the promise of deep decarbonization;
- as a systems thinker who understands that feedbacks are often more powerful than mandates;
- as a global citizen who believes that climate crisis is not just a technical challenge, but a systemic and societal imperative;
- as a father, colleague, and friend—deeply moved by the questions: *What kind of world are we leaving behind? What kind of ancestors are we becoming?*

To those working in policy, finance, urban planning, material science, energy, education, art, or agriculture—this book is for you.

To the activists, storytellers, researchers, and community builders—this book is for you.

To the students and youth wondering whether their agency matters—this book is for you most of all.

This book is also an invitation to those who feel overwhelmed or disillusioned.

It says: *You are not alone. And you are not small. Your coherence matters.*

In quantum systems, small perturbations can shift the entire field. That means your voice, your choices, your rituals, your relationships—they are not just responses. They are levers of transformation.

In that spirit, I thank you for walking alongside me through these pages.

If this book sparked a new thought, a new question, or a new commitment in you, I have done my work.

The rest, as ever, is entangled.

With gratitude,

Dr. Sanjay Wahal
Founder & President, Decarbonization, LLC
Former SVP, Technology Innovation, FyterTech Nonwovens
Author, *The eCarbon Card: Digital Solutions for Carbon Accountability*

Part I: Reframing the Crisis

Chapter 1

From Random Walks to Quantum Leaps

In 1973, economist Burton Malkiel published *A Random Walk Down Wall Street*[31], a radical rethinking of stock market behavior rooted in the idea that prices are unpredictable, chaotic, and governed by chance. At the time, this defied convention; investors preferred linear narratives—clear causes, rational actors, and forecastable outcomes. Half a century later, we face another system that defies predictability: **the planetary crisis**. Once again, linear models—emissions targets, GDP correlations, policy levers—are proving insufficient. The metaphor of a random walk captured a moment of epistemic humility in finance. What sustainability needs now is not randomness but quantum logic—an ability to accept simultaneity, paradox, entanglement, and emergent order. This book is built on that premise: we are trying to solve a nonlinear crisis with linear thinking, and it is falling short.

The Cognitive Trap of Linearity

Modern decision-making—especially in policy, business, and engineering—typically, clings to linear causality, where more X supposedly leads to more Y. Governments often assume that increasing electric vehicles automatically reduces emissions, but this ignores rebound effects, such as electricity sourced from coal or increased car usage. Similarly, rising urban temperatures are often addressed with tree-planting campaigns, which help but fall short without tackling urban sprawl, albedo, or water stress. Emissions are frequently attributed to fossil fuel use and population growth, yet this simplified Kaya identity neglects technology choices, economic structures, lifestyle shifts, and land-use dynamics. Countries like Bhutan, despite high per-capita energy needs, remain net carbon sinks thanks to forest cover and hydropower.

The climate system doesn't behave this way. Nor do supply chains, consumer behavior, or innovation networks. These are often nonlinear, dynamic systems where small interventions can yield large effects, as shown by the 1987 Montreal Protocol targeting CFCs, which significantly slowed ozone depletion and helped cool the planet. Feedback loops create reinforcing or dampening effects—Arctic ice melt, for instance, reduces Earth's albedo, causing more warming and ice loss in a dangerous cycle. Outcomes are highly sensitive to initial conditions; climate-resilient agriculture programs, for example, succeed or fail based on initial trust levels and localized knowledge, even when technical methods remain identical. Causality is distributed and delayed. Decades of coal burning in the Global North caused climate change, but its consequences (effects)—crop failures and flooding—are often borne by the Global South. Innovation networks also behave nonlinearly. Rooftop solar adoption followed a tipping curve: slow adoption led to social proof, which drove policy support and then rapid acceleration in some regions, bypassing others entirely.

We owe much of our progress to Newtonian thinking—linear models of cause and effect that have built engines, launched satellites, and optimized supply chains. These models work exceptionally well in structured, bounded, and low-variance systems where relationships are predictable and controllable. However, the climate crisis is neither structured nor bounded. It is shaped by feedback loops, tipping points, and emergent behaviors that defy straightforward prediction. While linear models remain indispensable for many aspects of planning—such as engineering renewable infrastructure or calculating emissions reductions—they are not sufficient on their own. We must complement them with nonlinear approaches that account for complexity, interdependence, and uncertainty. The real world is adaptive, time-sensitive, and interconnected, requiring tools that go beyond cost-benefit analyses, spreadsheets, or static policy memos. The challenge is not to abandon linear logic but to integrate it with systemic and dynamic models that can better navigate a landscape of entangled effects, hidden thresholds, and delayed responses.

As an engineer, I believe in numbers, models, and optimization, but I also know their limits. Climate is not a reactor, and society is not a flow diagram. In nonlinear systems, a good design flexes under pressure rather than working only under ideal conditions. Precision and uncertainty must dance together—gracefully.

Why Quantum Thinking?

In quantum mechanics, particles exist in superposition—multiple states at once. They are entangled, linked across space and time. Observing them changes their state. These are not just physical properties; they are metaphors for our sustainability dilemma. A policy can be both good and bad depending on the lens and stakeholder. Emissions in one country ripple across borders and decades. Measuring a system can change how its participants behave. To act wisely, we must let go of binaries, embrace uncertainty, and design for emergence rather than control. This is not abstract esoterica but a call to action: stop aiming for precision and start designing for resilience; stop optimizing individual parts and focus on the health of the entire system; stop measuring success solely by metrics and consider meaning and alignment as well.

The Architecture of This Book

This book is divided into six parts. The first, Reframing the Crisis, examines why linear models fail and introduces complexity theory, tipping points, and systems dynamics. The second, Quantum Thinking and Sustainability, unpacks metaphors from quantum theory and applies them to climate-relevant systems. The third, Solutions Beyond the Obvious, profiles out-of-the-box interventions, including behavioral policy tools, bio-based innovations, and digital governance. The fourth, Governance and Change, explores new paradigms in climate finance, regulation, and civic engagement. The fifth, Designing the Quantum Leap, proposes frameworks for leadership, innovation, and action under conditions of nonlinearity.

Finally, the sixth, Putting It All Together, synthesizes the conceptual, institutional, and technological threads into an integrated theory of change—translating nonlinear insight into practical strategy. It distills patterns across sectors, offers implementation pathways, and articulates what coherent action looks like when uncertainty is not a barrier but a design condition. Here, the book moves from exploration to orchestration: from isolated interventions to aligned systems, from insight to execution.

Throughout, I draw on my own experiences helping design, develop and commercialize green (biobased) materials—mycelium-derived alternate leather, seaweed-derived fibers, technical textiles, absorbent nonwovens, biopolymers etc., research on eCarbon Cards[32], carbon indices[33], and low-lignin gene-edited trees[34], as well as collaborations with governments, startups, investors, and academics. This book is both conceptual and practical. It challenges you to think differently while showing you what that might look like on the ground.

Who This Book Is For

This book is intended for policymakers tired of marginal gains, technologists seeking the next leap, systems thinkers looking for a new vocabulary, and citizens wondering where to begin. It offers a fresh approach for anyone who wants to make sense of and take action within the complexity of the climate crisis.

Let's Begin the Walk

We begin, as quantum walks do, not in a straight line but in multiple directions at once. Let go of your spreadsheet. Pick up your prism. Let's bend the light of sustainability through a different lens and walk into the nonlinear—together.

Chapter 2

The Failure of Linearity

The world behaves not like a machine but like a garden—alive, adaptive, sometimes unruly. And yet we persist in fixing it with blueprints and torque wrenches.

The Linearity Illusion

"In the presence of nonlinearity, ignorance of future inputs translates into ignorance of future outputs."

— *Nassim Nicholas Taleb, Antifragile (2012)*

Modern planning, policymaking, and engineering are built on the powerful yet not a correct assumption that *all* the systems behave linearly. This mindset underlies the tools we use to predict, optimize, and measure progress:

Emissions = Emission Factor × Activity Level

Efficiency = Output ÷ Input

ROI = Returns ÷ Investment.

Such equations work well in bounded, controlled systems like chemical reactors, assembly lines, or logistics chains—contexts where cause leads to effect in predictable, proportional ways. In these settings, more effort generally yields more progress.

But when systems are open, adaptive, and shaped by feedback—like Earth's climate, global trade, human behavior, or ecosystems—their behavior becomes associated with uncertainty and hence no longer predictable. These systems can remain stable for long periods, only to shift abruptly and dramatically when thresholds are crossed. Linear logic falters here, because the relationship between cause and effect is neither smooth nor proportional.

Over two centuries ago, Alexander von Humboldt[35, 36] described the natural world as a dynamic, interconnected system where climate, vegetation, soils, and human activity interacted in complex ways. In doing so, he anticipated what we now call nonlinear causality—feedbacks, thresholds, and cascading effects that defy linear prediction. His early ecosystem framing demonstrates that quantum or systemic thinking has deep intellectual roots, even if it was later overshadowed by industrial-era linear models. This book revisits that lineage, aligning Humboldt's vision with modern complexity science and sustainability practice.

Tipping Points and Their Role in Triggering Nonlinear Responses

Tipping points are thresholds beyond which systems undergo sudden, potentially irreversible change. In the climate system, they can be triggered by gradual shifts—such as rising average temperatures or altered rainfall patterns—until a critical limit is reached.

The following diagram (Fig. 2.1) illustrates the concept of tipping points:

Gradual Change → Threshold → Tipping Point → Abrupt Change → New State

Figure 2.1: Tipping Point

Climate Tipping Points: Nature's Nonlinear Warning

Three real-world phenomena reveal the limits of linear logic and highlight the importance of Tipping Points[37, 38]:

1. **Melting Arctic Sea Ice:** Reduces global climate stability by disrupting heat exchange patterns—which can lead to a tipping point in global climate regulation.
2. **Coral Reef Die-Offs:** Transform marine ecosystems and food webs—which can trigger a tipping point in marine ecosystems.
3. **Ice Sheet Collapse:** Accelerates sea-level rise beyond historical precedent—which can lead to a tipping point in sea-level rise.

Three real-world phenomena highlight the failure of linear logic and highlight the importance of Tipping Points.

Greenland Ice Sheet Melt

Rising global temperatures reduce the ice sheet's albedo (reflectivity), as soot deposition and melting expose darker surfaces that absorb more sunlight. This triggers a positive feedback loop: more melting → more heat absorption → faster melting. Once past a threshold, collapse becomes self-sustaining—no further warming required.

Coral Reef Bleaching

Corals rely on symbiotic algae for survival. Just a 1–2°C rise in ocean temperature can disrupt this delicate this relationship, triggering mass bleaching. The shift is abrupt, often irreversible on ecological timescales, and devastating to biodiversity.

Permafrost Thaw and Methane Release

When Arctic permafrost thaws, it releases methane—a greenhouse gas far more potent than CO_2—which accelerates further warming and thawing. This feedback could release millennia of stored carbon within decades, far exceeding what current models account for.

These systems exhibit nonlinearity in the most dramatic sense. These systems stay stable until they don't. The change is not gradual—it is threshold-driven, stepwise, and often catastrophic. We are not on a gentle slope of change—we are standing at a precipice.

Why Linear Models Still Dominate

"Precision in the wrong model is worse than uncertainty in the right one."

— *W. Brian Arthur, Complexity Economist*

Despite mounting evidence of nonlinearity, climate models, government policies, and economic instruments often rely on incremental measures such as reducing emissions by 2% per year, implementing a carbon tax of $50/ton to shift market behavior, or planting millions of trees to offset gigatons of CO_2. These approaches are essential and have played a critical role in advancing climate action, providing structure, accountability, and measurable progress. However, they are not sufficient on their own. While effective for many aspects of climate mitigation, these tools were designed for relatively stable and predictable systems. The climate crisis, by contrast, is a dynamic, nonlinear challenge characterized by feedback loops, tipping points, and cascading risks. What we need is not to abandon these strategies but to complement them with adaptive, systems-level approaches that can anticipate and respond to complexity, scaling both immediate actions and transformative solutions simultaneously.

The limitations of our tools further compound this problem. As Nobel laureate Elinor Ostrom observed,

"The tools we use shape our thinking. Models that assume equilibrium and linear response train us to see the world that way, even when the world doesn't comply."

Climate action is not limited by data but by overreliance on outdated tools—cost-benefit analyses that discount future generations, carbon accounting models that treat social systems as static, and optimization algorithms that reward local efficiency while ignoring global failures. Donella Meadows famously reminded us:

"We cannot impose our will on a system. We can listen to what the system tells us, and discover how its properties and our values can work together to bring forth something much better than could ever be produced by our will alone."

Cognitive bias adds another layer of challenge. As Geoff Mulgan noted in *Big Mind* (2017)[39],

"What makes systems nonlinear is that the effect is rarely proportional to the cause. The real world is much messier than our spreadsheets suggest."

Our brains, built for survival on the savanna, prioritize short-term threats and linear causality, leaving us ill-equipped for slow, system-wide crises like climate change. We fall prey to status quo bias, preferring inaction or incremental steps even amid deep risks; the availability heuristic, which makes vivid events overshadow complex trends; confirmation bias, which filters information through preexisting beliefs; and hyperbolic discounting, which undervalues long-term benefits and harms, leading to chronic underinvestment in resilience. As Fritjof Capra observed,

"The major problems in the world are the result of the difference between how nature works and the way people think."

Consequences of the Linearity Trap

Ignoring system dynamics leads to underestimated risks, overpromised benefits, and misallocated resources. Early IPCC reports underpredicted sea-level rise and Arctic melt due to weak modeling of feedbacks[40]. Carbon offset schemes assume a tree planted today can offset fossil emissions lasting centuries—an oversimplification that undermines urgency.

In systems with tipping points, early, front-loaded action is essential. Waiting for perfect clarity only increases the risk of abrupt, irreversible shifts.

Linear Solutions for Nonlinear Problems: A Systemic Mismatch

Linear measures often collide with nonlinear realities.

- **Energy Efficiency and EV Adoption** – Efficiency gains and cheaper EVs can lead to the *rebound effect*, increasing total energy use. While sometimes offsetting emissions reductions, in contexts like China, rapid EV adoption has accelerated decarbonization without major market distortion—especially when paired with grid decarbonization.
- **Plastic Bans** – Reducing single-use plastics cuts visible waste, but shifting to paper or bioplastics can raise water and energy demand unless life-cycle impacts are considered.
- **Transport Infrastructure** – Building new roads can reduce congestion short-term but may induce more car use (*Braess's Paradox*)[41] unless balanced with public transit and active mobility infrastructure.

- **Irrigation for Drought Relief** – Boosts crop yields in the short term but can cause aquifer depletion and soil salinization without sustainable water management.

These examples underline the need to pair interventions with systemic strategies that anticipate side effects and reinforce long-term resilience.

Lessons from Complexity Science

The climate crisis is a **complex adaptive system (CAS)**—characterized by feedback loops, thresholds, emergence, delayed causality, and non-proportionality:

- **Feedback Loops**[42, 43, 44, 45] – Self-reinforcing processes or cycles which can amplify (warming → melting → more warming) or dampen - stabilizing loops (forest regrowth → CO_2 absorption) climate responses.
- **Thresholds**[46, 47, 48, 49, 50] – Points beyond which change becomes irreversible.
- **Emergence**[51, 52] – New patterns and behaviors arising from interactions among system components and not present in in individual components.
- **Delayed Causality**[53, 54, 55] – Today's actions may have impacts decades later, making attribution challenging.
- **Non-Proportionality**[56] – Small changes can cause outsized disruptions e.g. 1% change can trigger 100% disruption.

To navigate such a system, we must think in systems rather than silos, model dynamics instead of static snapshots, and design interventions that are adaptive rather than fixed. We must embrace uncertainty rather than attempt to control it and prioritize resilience over narrow optimization.

Feedback Loops in Practice

Feedback loops are self-reinforcing processes that can either amplify (*positive feedbacks*) or dampen (*negative feedbacks*) climate responses. Positive feedback loops can accelerate climate change, while negative feedback loops can slow it down. Some examples of feedback loops include:

Positive

- **Water vapor feedback loop** – As global temperatures rise, evaporation increases, adding more water vapor to the atmosphere. Because water vapor is a potent greenhouse gas, this additional moisture traps more heat, further accelerating warming.
- **Ice-albedo feedback loop** – When ice cover decrcases, less sunlight is reflected back into space and more is absorbed by the darker underlying ocean or land. This extra heat accelerates ice melt, reinforcing the initial warming.

Negative

- **Vegetation regrowth (CO_2 fertilization) feedback** – In some regions, rising atmospheric CO_2 concentrations stimulate plant growth, allowing vegetation to absorb more carbon from the atmosphere. This can slow the rate of warming, although the effect is constrained by nutrient availability, water stress, and ecosystem degradation.
- **Cloud formation feedback** – In certain conditions, warmer surface temperatures increase evaporation, leading to more low-level cloud formation. These clouds reflect a greater share of incoming solar radiation (raising planetary albedo), which can cool the surface and partially offset the initial warming.

The following ordinary differential equation (ODE)[57,58] represents the feedback loop process:

$$\frac{dT}{dt} = \lambda T + \mu F(T)$$

where T is the temerature, λ is the feedback parameter, and F(T) is the feedback function.

This is a nice "template" for simple climate-feedback models: $T(t)$ is global-mean temperature anomaly, λ is the "base" (un)stabilized tendency, and $\mu F(T)$ is feedback (Planck, ice–albedo, clouds, etc.). Below are three concrete, climate-flavored choices of F(T) with closed-form solutions provided as an illustration of sensitivity of parameters to global warming.

<u>Linear (Planck + Restoring) Feedback</u>

A very common first model assumes the net feedback is proportional to how far we are from a reference T∗ (e.g., planetary radiative restoring toward equilibrium):

$$F(T) = -(T - T_*)$$

Then

$$\frac{dT}{dt} = \lambda T + \mu(-(T - T_*)) = (\lambda - \mu)T + \mu T_*$$

Solution (linear ODE with constant forcing)

Let $k = \lambda - \mu$. For $k \neq 0$,

$$T(t) = T_{ss} + (T(0) - T_{ss})e^{kt}, \qquad T_{ss} = \frac{\mu}{\mu - \lambda} T_*$$

- If $\mu > \lambda$ (restoring dominates,) then $k < 0$ and $T(t) \to T_{ss}$ exponentionally with e-fold timescale $\tau = \frac{1}{\mu - \lambda}$.
- Interpretation (global warming step): If T_* represents the new equilibrium set by a step increase in external forcing (e.g., CO_2 doubling), then T_{ss} is the equilibrium warming and the trajectory is a classic "fast approach to a new steady state."

Numerical example. Say $\lambda = 0.2\ \ yr^{-1}, \quad \mu = 0.5\ \ yr^{-1}, \quad T_* = 3^\circ C, \quad T(0) = 0.$

Then $k = -0.3, \quad \tau \approx 3.33\ \ yr, \quad T_{ss} = \frac{0.5}{0.5 - 0.2} \cdot 3 \approx 5^\circ C.$

So $T(t) = 5\left[1 - e^{-0.3t}\right]{}^\circ C.$

For example, for t = 10 yr, T(10) ≈ 4.75 °C

<u>Constant External Forcing + Linear Planck Feedback</u>

Energy-balance models are often written (after dividing by heat capacity) as

$$\frac{dT}{dt} = -\alpha T + \beta \Delta F,$$

where $\alpha > 0$ is the (stabilizing) Planck feedback and ΔF is an imposed radiative forcing (e.g., a CO_2 step). This matches the template with $\lambda = -\alpha, \quad \mu = \beta, \quad F(T) \equiv \Delta F$ (a constant)

Solution: $T\left(t\right) = T_{eq} + (T(0) - T_{eq})e^{-\alpha t}, \ T_{eq} = \frac{\beta}{\alpha} \Delta F.$

- If ΔF jumps from 0 to 3.7 W m^{-2} (canonical CO_2 doubling), and if $\frac{\beta}{\alpha} \approx 0.8\ {}^\circ C/\left(W\ m^{-2}\right)$, then $T_{eq} \approx 3^\circ C$. The system relaxes exponentially to ~3 °C warming with e-fold 1/α (or τ) (years to decades depending on ocean heat capacity details).

- If ΔF jumps from 0 to 3.7 W m^{-2} (canonical CO_2 doubling), and if $\alpha = 0.25\ \text{yr}^{-1}$, $\beta = 0.8$ °C yr^{-1}/(W m^{-2}), then

$$T_{eq} = 11.84\ °C \text{ and } \tau = 4.0\ \text{yr}$$

(note: this large equilibrium reflects the chosen $\beta/\alpha = 3.2$. If one wants an ECS $\approx$ 3°C , set $\beta/\alpha \approx 0.8$ °C /(Wm^{-2}) as we did above; but if you prefer a smaller response factor, we can tune it as shown below.)

Ice-Albedo-Type Nonlinear Positive Feedback

To capture the idea that warming amplifies further warming (e.g., reduced ice → lower albedo → more absorbed sunlight), a simple caricature is a cubic nonlinearity:

$$\frac{dT}{dt} = \lambda T - \kappa T^3, \quad (\kappa > 0).$$

This is the normal form for growth limited by strong positive feedback at larger T. It fits the template with $F(T) = -T^3$ and $\mu = \kappa$ (taking $\lambda > 0$) for net growth at small T).

Equilibria and stability

$$0 = \lambda T - \kappa T^3 \quad \Rightarrow \quad T^* \in \left\{0, \pm\sqrt{\frac{\lambda}{\kappa}}\right\}.$$

- $T^* = 0$ is unstable if $\lambda > 0$ (any small warming grows).
- $T^* = \pm\sqrt{\frac{\lambda}{\kappa}}$ are stable (system saturates at a warm or cool state).

Closed-form solution (for $T(0) \neq 0$).

Define $T_{eq} = \sqrt{\frac{\lambda}{\kappa}}$ and $r = \lambda$. For $T(0) > 0$,

$$T(t) = \frac{T_{eq}}{\sqrt{1 + \left[\left(\frac{T_{eq}}{T(0)}\right)^2 - 1\right] e^{-2rt}}} = \frac{T_{eq}}{\sqrt{1 + Ae^{-2\lambda t}}}, A = \left(\frac{T_{eq}}{T(0)}\right)^2 - 1$$

(For $T(0) < 0$ the solution approaches $-T_{eq}$ with the same functional form.)

Interpretation. Small anomalies grow (positive feedback), but the cubic term limits runaway, producing a finite warm equilibrium—qualitatively like ice-albedo feedback saturating as ice largely disappears.

Numerical example.

Let $\lambda = 0.1\ \text{yr}^{-1}$, $\kappa = 0.01\ ^\circ C^{-2} \text{yr}^{-1} \Rightarrow T_{eq} = \sqrt{10} \approx 3.162^\circ C$. If $T(0) = 0.5^\circ C$

$A = \left(\frac{3.162}{0.5}\right)^2 - 1 \approx 39$. Then

$$T\left(t\right) = \frac{3.162}{\sqrt{1+39e^{-0.2t}}}\ ^{\circ}C,$$

Which rises sigmoidal-like toward $\approx \mathbf{3.16}^{\circ}\mathrm{C}$ with $\tau \approx 10$ yr.

How to read this framework

- **Choose** F(T) to represent the feedback you care about (linear restoring, constant external forcing, nonlinear ice-albedo, saturating cloud feedbacks, etc.)
- **Solve/interpret** the resulting ODE: equilibria from $\lambda T + \mu F(T) = 0$; stability from $\lambda + \mu F'(T)$ evaluated at equilibria; time scale from the linearized growth/decay rate.

Two other useful *F(T)* choices (qualitative, typically solved numerically)

- Saturating feedback: $F(T) = \tanh(T/T_s)$. Linearizing near 0 gives an effective growth rate $\lambda + \mu/T_{si}$; **for large** $|T|$, $F \rightarrow \pm 1$, so feedback caps out, preventing divergence.
- **Water-vapor + cloud anelesticity:** sometimes approximated as $F\left(T\right) = aT + bT^3$ to combine linear and superlinear pieces; you get the same analytical tools as in example 3, with thresholds/ bistability possible.

<u>Threshold Effects in Intervention Design</u>

Threshold effects play a critical role in shaping sustainability intervention design because they highlight how systems can tolerate gradual change up to a point, beyond which small additional stresses can trigger abrupt, large-scale, and often irreversible shifts. This understanding fundamentally influences how interventions are planned and prioritized:

- **Precautionary Approach** – Knowing that ecosystems and climate systems have thresholds (or tipping points) means interventions must be precautionary—acting before these critical points are approached. Delaying action risks crossing thresholds that make recovery difficult or impossible.
- **Early, Strategic Action** – Interventions are designed to be proactive where possible, focusing on preventing system stress from

reaching dangerous levels. Monitoring and rapid response mechanisms are built in to detect early signs of approaching thresholds, so interventions can be deployed before irreversible damage occurs.

- **Design for Resilience[59, 60, 61], Not Just Efficiency** – Threshold effects mean that once a system crosses into an undesirable state, incremental improvements might not restore it. As a result, sustainability interventions emphasize resilience (capacity to absorb shocks and reorganize), not just technical efficiency, to guard against abrupt transitions.
- **Targeted Intervention at High-Leverage Points[62]** – Threshold effects often identify critical nodes or processes in a system (such as key ecosystems or feedback loops) where small, focused actions can have disproportionately large protective effects. Interventions are designed to maximize impact at these leverage points to prevent crossing system thresholds.
- **Adaptive Management and Flexibility** – Because the exact location of thresholds in complex systems is often uncertain, interventions are adaptive, with ongoing monitoring and willingness to revise approaches if signs indicate the system is close to a threshold.
- **Holistic and System-Wide Perspective** – Threshold effects encourage designers to look beyond single variables and consider the overall health and structure of systems, ensuring interventions do not inadvertently push parts of the system past critical limits.

Example: In water management, understanding that a lake may suddenly shift from a pristine to a turbid state if nutrient loading crosses a threshold leads to strict early controls, rapid detection systems, and buffer strategies that keep nutrient inputs below safe levels.

In sum, threshold effects prompt intervention design that is cautious, resilient-focused, adaptive, and prioritized around preventing abrupt systemic collapse, rather than waiting for gradual deterioration before acting. This leads to more robust, forward-looking, and effective sustainability strategies.

Emergence

From zebra stripes to self-organizing cities, emergence shows how complex structures arise without central control. In nature, ant colonies construct intricate tunnel networks without any master plan, guided only by local interactions and pheromone trails. Flocking starlings create vast, shifting murmurations through simple alignment and spacing rules followed by each bird. In human systems, informal settlements often develop their own street layouts, markets, and social governance without formal urban planning, adapting dynamically to population needs. In climate systems, new dynamics—such as altered ocean currents in the North Atlantic that disrupt weather patterns across continents—can emerge suddenly and reshape entire regions. Similarly, rapid shifts in vegetation cover, like the greening of Arctic tundra, can emerge from local warming patterns, altering carbon uptake and regional climate feedbacks in ways not directly orchestrated by any single driver.

Delayed Causality

Because climate systems have long lag times, emissions today can commit us to changes decades later. For example, CO_2 emitted from fossil fuel burning can persist in the atmosphere for centuries, meaning today's energy choices will still be influencing global temperatures in the 22nd Century. The thermal inertia of the oceans delays the full expression of warming—heat absorbed now may not fully affect weather patterns for decades, yet it locks in future sea-level rise from thermal expansion and ice sheet melt. Similarly, deforestation today can reduce regional rainfall over decades, with agricultural decline appearing long after the trees are gone. This delay makes it tempting to postpone action—but also means proactive measures, such as rapidly deploying renewables or restoring ecosystems, have profound, long-term payoffs that will be felt for generations.

Non-Proportionality

In nonlinear systems, effects are not proportional to causes. A 1–2 °C increase in average ocean temperature—seemingly small—can trigger mass coral bleaching and collapse of entire reef ecosystems, while much larger swings in other variables may have only minor effects. Melting just a few percent of Arctic summer sea ice can sharply alter jet stream behavior, leading to persistent droughts or floods thousands of kilometers away. Conversely, massive investments in carbon capture deployed in the wrong locations or at the wrong scale may yield negligible climate benefit if they fail to address the main sources of emissions or interact with existing feedback loops. In short, small disturbances at the wrong thresholds can be catastrophic, while large efforts applied without systemic insight can achieve surprisingly little.

A New Vocabulary for a New Age

Addressing nonlinear crises requires a shift in language and thought:

- **Tipping Points** – Critical thresholds triggering irreversible change (not just inflection points).
- **Phase Transitions** – Rapid shifts between system states (not just market transitions).
- **Hysteresis** – Systems remaining in a changed state even after the cause is removed (not just elasticity).
- **Coevolution** – Mutual adaptation of interconnected systems (not just independence).
- **Superposition** – Holding multiple possibilities until observation resolves outcomes (not just single-variable causality).

We must speak in the language of living, adaptive systems, not just mechanical metaphors. These terms reflect the underlying dynamics of living, interconnected systems and highlight how deeply our current framing falls short.

Prelude to the Walk Ahead

This chapter serves as a door.

To step through it is to accept a simple but profound truth:

while spreadsheets, models, and forecasts are indispensable tools in managing the climate crisis, they cannot solve it alone. The challenges we face are not purely technical or linear—they involve complex interactions across ecological, economic, and social systems that resist simple modeling. Spreadsheets help us measure, plan, and track progress, but they must be complemented by approaches that embrace uncertainty, interconnectedness, and systemic feedbacks. Solving a planetary crisis requires both the precision of analytics and the adaptability of nonlinear thinking.

We must learn to walk differently.

To see sideways.

To act when the path is unclear—not because we are reckless, but because waiting for clarity is the greater risk.

In the next chapter, we will enter this new space of thinking—one where probability, paradox, and entanglement replace certainty, hierarchy, and linearity.

Let us begin the quantum walk.

Chapter 3

The Climate Crisis as a Complex Adaptive System (CAS)

Trying to solve the climate crisis with a linear plan is like trying to play chess using only checkers' rules.

"We Need to See the Whole Board to Stop Climate Change."
— Dr. Jonathan Foley

The climate crisis is often reduced to an emissions problem, but this framing is far too narrow. What we are truly facing is the breakdown of planetary homeostasis—an unraveling of tightly coupled systems that, until now, have regulated life-supporting conditions on Earth. The climate is much more than a thermostat that can be adjusted by turning a single dial. It is a complex adaptive system (CAS), defined by nonlinear feedbacks, delayed effects, and emergent dynamics. Sustainability, like climate, is not static but evolves through dynamic interplay between people, policies, technologies, ecosystems, and values. Recognizing this interconnectedness allows us to move beyond rigid, top-down plans toward systems that can learn, adapt, and regenerate.

Sustainability is not just a checklist of actions but a multi-scalar, multi-agent process shaped by ecological feedbacks such as soil regeneration or biodiversity loss; economic decisions on consumption, investment, and innovation; cultural narratives and behaviors, including consumerism or sufficiency; policy and institutional dynamics like regulations, subsidies, and norms; technological disruptions ranging from artificial intelligence to renewable energy or gene-edited crops; and social learning through grassroots innovations and indigenous knowledge. These elements are not separate—they are interconnected and dynamic, exhibiting the hallmarks of CAS: interdependence where actions in one domain ripple across others; dynamism as goals and thresholds shift over time; emergence as unexpected patterns arise; path dependence where outcomes hinge on historical conditions; and self-organization where change often originates at the margins, not from centralized control.

Linear planning models often fall short because they underestimate these dynamics. True sustainability demands approaches that are iterative, reflexive, and responsive—geared not toward rigid control but toward resilience, co-evolution, and generative potential.

In the rest of this chapter, we will explore key features of complex systems—such as feedback loops, tipping points, cascading failures, and regime shifts[63]—and illustrate their relevance not only to climate science but to all domains of sustainability thinking.

What Is a Complex Adaptive System?

Before we can reimagine solutions, we must understand the nature of the challenge. The climate crisis is not just an environmental issue; it is a multi-scale, multi-actor, multi-feedback system—a true CAS.

A CAS has five defining characteristics:

1. **Interconnectedness:** Its components are interdependent; a change in one affects many others.
2. **Emergence:** It produces system-wide emergent behaviors that cannot be predicted by analyzing individual parts in isolation.
3. **Feedback Loops:** Actions trigger reactions, which amplify or dampen the original input.
4. **Adaptation:** The system adapts and reorganizes in response to pressures.
5. **Nonlinearity (Non-Proportionality):** Outputs are not proportional to inputs, exhibiting linearity; small nudges can trigger large shifts—and vice versa.

Climate, economy, ecology, and human behavior all share these properties. Addressing climate change is thus not a task of optimization; it is a challenge of navigation.

Earth's Climate: The Archetype of a CAS

The Earth's climate system consists of interacting subsystems—the atmosphere, oceans, cryosphere, biosphere, geosphere, and anthroposphere (human systems). Each has inputs, such as solar radiation or greenhouse gas emissions; internal dynamics, like cloud formation or ocean currents; feedback loops, including the carbon cycle; and interactions with other subsystems. Warmer oceans, for instance, reduce CO_2 solubility, which in turn amplifies warming. Deforestation diminishes evapotranspiration, altering rainfall patterns and affecting agriculture and migration, which subsequently shape policy responses. Predicting a single linear outcome from climate interventions is nearly impossible because each action cascades through a web of interconnected subsystems.

A Sustainability Systems: The Extended CAS Frame

Climate is not the only CAS. Sustainability across energy, food, water, biodiversity, economy, equity, and institutions* is equally governed by CAS principles. These systems co-evolve, resist top-down control, and exhibit behaviors such as emergence, where informal slum economies self-organize around water scarcity; path dependence, where fossil fuel subsidies lock societies into high-carbon pathways; co-evolution, as consumer demand for sustainable products reshapes business models; self-organization, exemplified by community-owned renewable energy projects; and feedback loops, like water scarcity driving migration and urban stress, leading to political tension. Nonlinear effects, such as coral reef bleaching that collapses fisheries and food security, highlight how these systems operate outside linear expectations.

*These categories represent a domain-specific lens within the broader SCI framework. They map directly onto the SCI framework: ecological integrity covers biodiversity and ecosystem health, planetary boundaries encompass energy, water, and food systems, and social equity and institutional resilience connect to equity and governance structures. For full alignment with SCI, see Appendix XVII Table XVII.2.

Why This Matters

Understanding sustainability as a **network of complex adaptive systems (CAS)** shifts the frame of intervention. It shows why routine interventions or reductionist actions—such as planting trees without changing land-use incentives or subsidizing solar panels without grid reform—often **backfire** or fail to **scale**. In such systems, interventions ripple through feedback loops and produce unintended consequences.

The critical question thus changes from:

"What is the right solution?" → (a linear search for a silver bullet) to *"How do we design systems that can learn, evolve, and stay resilient?"*

Examples of this reframing in practice:

- **Agriculture** – Rather than pushing a single "climate-smart crop," adaptive farming systems blend agroecology, diversified cropping, digital monitoring, and community-based seed networks.
- **Urban Mobility** – Instead of merely deploying EV buses, a learning system designs flexible transit ecosystems—integrating on-demand services, bike-sharing, and dynamic pricing that adapts to rider behavior.
- **Energy** – Rather than installing more solar panels in isolation, resilient energy systems are designed as *adaptive microgrids* that can rebalance, self-heal, and evolve as demand and technology shift.
- **Waste Management** – Instead of building more landfills (a single-point "solution"), circular systems design material flows for reuse, repair, and regeneration.

The message being that **no intervention exists in isolation**. The real work is creating conditions for *continuous learning, feedback, and co-evolution*—systems that adapt as reality changes, just as ecosystems have always done.

Examples of CAS Behavior in Climate Dynamics

CAS behaviors manifest vividly in climate dynamics. *Emergence* examples are common. Megadroughts, for example, do not have a single cause but emerge from overlapping factors such as long-term warming, short-term anomalies, soil moisture loss, and deforestation-induced rainfall changes. Urban heat islands, which trap heat in concrete and asphalt, lead to increased cooling demands and strain electricity grids during heatwaves, creating blackouts that are not the result of one actor's decision but systemic interaction. *Path dependence* is seen in Arctic sea ice loss, where reflective ice replaced by dark ocean triggers self-sustaining warming even if emissions are curbed. Similarly, permafrost thaw releases methane locked in for millennia, amplifying warming due to historical emissions—a trajectory that early action could have mitigated but which is now partially locked in. The South Asian monsoon illustrates *adaptation* and *co-evolution*: ocean warming, land-use change, and aerosols have altered rainfall patterns, forcing farmers to shift crops and irrigation strategies while economic and social systems adapt in uneven and often mismatched ways. The Atlantic Meridional Overturning Circulation (AMOC), often simplified as the Gulf Stream, is slowing due to freshwater influx from melting ice. This alters weather patterns across Europe and Africa. These shifts, in turn, trigger **economic, migratory, and political responses**—a human adaptation to physical system change.

Human Systems as CAS

Human systems—governance, culture, and economics—are themselves CAS embedded within sustainability. Governance functions like an ecosystem, adapting to shocks such as protests, pandemics, or elections. Social norms, like the adoption of plant-based diets or solar power, often shift abruptly after a tipping point of social proof is reached. Markets evolve in emergent ways, as seen in ESG investing or the rise of carbon pricing. Cultural narratives, such as consumerism or sufficiency, influence what policies are possible and which behaviors gain legitimacy. We are both the agents of transformation and part of the terrain that must shift. Without adaptive institutions and cultural narratives, technical solutions alone will not scale.

Implication for Sustainability Transitions

Understanding human systems as CAS underscores a core paradox: We are both the agents of transformation and the terrain that must shift. Top-down interventions alone will fail without nurturing the adaptive capacity of institutions, the reflexivity of policy, and the emergence of new cultural narratives. Leverage comes not just from regulating emissions, but from modifying the system's capacity to learn, adapt, and self-correct.

This framing sets the stage for later chapters on institutional design, quantum-aligned governance, and co-leveraged interventions across domains. Human systems are not just levers; they are leverage points in themselves—if understood as adaptive, relational, and capable of transformation.

The Fallacy of Reductionism

Reductionist thinking—breaking complex problems into isolated parts—falls short in CAS. Here problems are interwoven. Solutions must leverage **cross-sectoral synergies** to avoid backfiring. Soil health, for instance, is not just an agricultural issue but affects carbon sequestration, water cycles, and biodiversity. Urban transport solutions, such as building more roads, often induce more car use rather than alleviating congestion, a dynamic known as Braess's Paradox[64]. Plastic bans, when not approached systemically, can shift harm to other materials. Likewise, interventions like clean cooking stoves reduce deforestation and black carbon while empowering women, and school food programs link local agroecology with improved nutrition, lower logistics emissions, and long-term dietary changes. Such examples illustrate the importance of integrated approaches that acknowledge the interdependence of systems.

Phase Transitions and Critical Thresholds

Having accumulated hidden stresses, CASs often undergo abrupt shifts, once critical thresholds are crossed—much like water boiling or ice melting. Examples include Arctic Sea ice loss with irreversible albedo shifts, the collapse of coral reefs after successive bleaching events, disruption of the Atlantic Meridional Overturning Circulation (AMOC) leading to abrupt weather changes, monsoon shifts that affect billions, Amazon dieback converting rainforest to savanna due to deforestation and drought, simultaneous crop failures causing food system collapse, and methane release from permafrost accelerating global warming. These events underscore the urgency of proactive action before systems reach tipping points.

Designing in the Face of Complexity

To work within CAS or designing for CAS, we must:

- **Embrace feedback** – Monitoring and learning loops.
- **Design for emergence** – Create enabling conditions, not rigid plans.
- **Prioritize resilience** – Flexibility over efficiency.
- **Use scenario thinking** – Prepare for multiple futures.
- **Empower self-organization** – Enable local adaptation.

In this sense, we must act less like engineers seeking control and more like **choreographers** or **conductors**—balancing structure with improvisation and knowing when to lead or guide and when to let systems self-organize or to "hold space" for adaptive, distributed action.

Complexity Is Not Chaos—It's Opportunity

Complexity, while daunting, offers opportunities for transformation. Small, well-placed interventions can cascade into large systemic changes. Local innovations like agroecology or microgrids can scale globally when they align with cultural and market tipping points. Cultural shifts often precede and accelerate policy, making collective narratives powerful drivers of change.

Complex systems are not hopeless. They offer:

- multiple pathways for intervention
- leverage points where small changes cascade
- the ability for local solutions to scale through tipping points

Why This Matters Now

The climate emergency is accelerating, and we are entering a critical window where social, ecological, and technological systems are primed for nonlinear transformation. The risk of inaction is not gradual decline but abrupt, irreversible change. To seize this moment, we must adopt quantum-aligned strategies—ones that embrace entanglement, uncertainty, and emergence.

Transition to the Quantum Mindset

The next chapter introduces quantum thinking, exploring how concepts like superposition, entanglement, and observer effects offer fresh metaphors and strategies for sustainability. Superposition encourages us to hold multiple possibilities simultaneously, entanglement reveals the deep interconnections across systems, and observation reminds us that how we measure, and act changes the system itself. With these principles, we step into the quantum domain.

Part II: Quantum Thinking and Sustainability

Chapter 4

What Is a Quantum Walk?

Embracing Superposition, Entanglement, and Uncertainty for Systemic Sustainability

In a quantum walk, we don't choose a path—we explore all paths at once, and the future selects what survives.

Walking Differently: From Newton to Schrödinger

We walk the way we think—linearly, step by step. This has been Newton's world: predictable, mechanical, and deterministic. It has given us engines, factories, and forecasts, all based on the logic that if you push the system, it moves accordingly. We extended this mechanistic worldview to economics, governance, and even the planet. Yet quantum mechanics now challenges this framework, showing that in the quantum world:

- particles exist in multiple states until measured (*superposition*)
- their fates are interdependent (*entanglement*)
- observing them alters their state (*observer effect*)
- probability replaces certainty
- time flows strangely—nonlinear and non-deterministic

To walk quantumly is to embrace complexity—not by seeking control but by leaning into curiosity and uncertainty. It is to operate in a **probabilistic, emergent, and indeterminate** world. This is not chaos. It is *possibility*, and it offers a design principle for systemic sustainability solutions.

This chapter explores how the quantum walk offers not just a metaphor but a design principle for systemic sustainability solutions.

What Is a Quantum Walk?

A **quantum walk** is the quantum analogue of a classical random walk.

In a **random walk** (e.g., stock prices or Brownian motion), a walker moves step-by-step, each direction chosen randomly.

In a **quantum walk**:

- The walker explores **all possible paths simultaneously**.
- Each path has a **probability amplitude**, not a binary outcome.
- Paths **interfere** with each other—amplifying or canceling.
- **Entangled states** allow distributed yet coherent behavior.

Two Core Features

Superposition – A quantum walker exists in multiple locations at once. Instead of asking, "*Where is it now?*", we ask, "*What is the probability distribution of where it could be?*"

Entanglement – Particles (or agents) become linked so that the state of one instantly influences the other, regardless of distance. This creates distributed intelligence, not centralized control.

From Physics to Metaphor: Why This Matters for Sustainability

Quantum concepts offer deep metaphors for sustainability. Superposition teaches us to hold multiple pathways or solutions simultaneously rather than prematurely collapsing to a single option. Entanglement reminds us of the interdependence of energy, economy, culture, and climate. The observer effect shows that metrics and framing reshape reality, as with GDP versus measures of well-being. Interference illustrates how policies can amplify (*constructive*) or cancel (*destructive*) each other like overlapping waves. Probability urges us to design for resilience and likelihoods rather than pursuing one "optimal" outcome, and collapse reminds us that taking action reveals one reality, but collapsing choices too early narrows our potential.

A quantum walk is about disciplined exploration of multiple futures, using *strategic ambiguity* and *multi-dimensional design*, rather than rushing to singular solutions.

Sustainability Demands Superposition

In sustainability, we often ask:

- *"Should we prioritize adaptation or mitigation?"*
- *"Should we invest in hydrogen or electrification?"*
- "*Is nuclear the savior or the threat?*"

These binary (*either/or*) frames miss the point. The real answer is often both/and solutions.

Superposition in Action – Key Examples

Superposition reflects the ability of systems to hold multiple solutions simultaneously, enabling flexibility, resilience, and accelerated decarbonization without waiting for a single "perfect" pathway. Across sectors, leading initiatives demonstrate how layered strategies create compounded benefits.

Urban Climate Design

Superposition: Cities like Barcelona (Superblocks) and Singapore combine dense housing with vertical greenery, bike highways, and stormwater parks.

Impact: Reduces emissions, mitigates heat islands, improves livability, and supports biodiversity.

Adaptive Building Materials

Superposition: Electrochromic windows (e.g., SageGlass retrofits at Willis Tower, Chicago) dynamically tint to reduce HVAC loads.

Impact: 20–30% cooling energy reduction, improved comfort, no trade-off between daylight and energy.

Decarbonizing Cement

Superposition: LC3 cements (Limestone + Calcined Clay) combined with CCUS retrofits in plants like Heidelberg Cement (Norway).

Outcome: >40% CO_2 reduction, maintaining performance while diversifying pathways.

Green Hydrogen & Electrification in Steelmaking

Superposition: HYBRIT (Sweden) uses both hydrogen-based DRI and electric arc furnaces.

Impact: Flexible decarbonization before full hydrogen scaling.

Bio-feedstocks & Circular Chemistry

Superposition: Companies like BASF blend biomass-based naphtha with chemical recycling.

Outcome: Keeps carbon in circulation and reduces fossil demand.

Mobility Futures

Superposition: Amsterdam's Zero-Emission Mobility Plan integrates public transit, EV fleets, bike highways, and logistics electrification.

Impact: Seamless hybrid mobility, reducing both congestion and emissions.

<u>Systems Insight</u>

These examples illustrate that sustainability thrives on holding multiple pathways open—superposition in action.

Entanglement: The New Interdependence

The climate crisis is **hyper-entangled**. Actions in one domain ripple across others:

<u>Entanglement Examples</u>

Entanglement illustrates how climate, economic, and social systems interact across geographies and sectors. Actions in one domain frequently trigger cascading consequences elsewhere, reinforcing the need for solutions designed around interconnected outcomes rather than isolated interventions.

Cross-Regional and Global Entanglement

- Stubble burning in Punjab → Delhi's AQI → Urban health crises.
- Wildfires in Canada → Air quality decline in U.S. cities (e.g. in upper midwestern states) → Healthcare costs.
- Ocean warming → Fishery collapse → Trade disputes → Food insecurity.
- Consumer demand in L.A. → Amazon deforestation → Indigenous land conflicts.

Solution-Driven Entanglement (Positive Cascades)

- **Solar Microgrids in Rural India** – Simultaneously expand energy access, empower women, reduce kerosene pollution, and improve educational outcomes.
- **Urban Tree Planting & Stormwater Design** – Trees mitigate heat islands while absorbing runoff, reducing flood risk.
- **Carbon Pricing with Dividends (British Columbia)** – Aligns market signals with social equity through revenue redistribution.
- **Regenerative Agriculture** – Generates co-benefits across soil carbon, biodiversity, rural livelihoods, and food sovereignty.
- **Circular Textile Systems** – Recycled fibers connect waste reduction, local job creation, and water conservation.

<u>Systems Insight</u>

Entanglement reveals that effective climate and sustainability strategies must be designed as multi-benefit interventions — addressing environmental, economic, and societal dimensions simultaneously rather than optimizing for a single metric.

The Observer Effect in Metrics and Mindsets

"When a measure becomes a target, it ceases to be a good measure."
— Marilyn Strathern.

Metrics are not neutral; they shape behaviors and incentives. ESG ratings can drive firms to game disclosure frameworks instead of reducing actual impacts. Forest carbon offsets, when mismanaged, can result in monoculture plantations that prioritize carbon storage over biodiversity. EV sales metrics may encourage higher car ownership without addressing congestion or sprawl. GDP growth still values extraction over regeneration. Smart city sensors, while offering data, can alter behavior in ways that erode trust. The lesson is that metrics must be designed carefully, balancing measurement with meaning and context.

Lesson: Metrics are not neutral; they shape system behavior and incentives.

Interference and Policy Design

Policies interfere like waves—sometimes amplifying, sometimes canceling each other.

Germany's Energiewende achieved *constructive* interference by pairing feed-in tariffs with job mandates, scaling both renewables and local economies. Norway's EV adoption, now at around 80%, was fueled by EV subsidies combined with a hydropowered grid. Indigenous land rights paired with forest payments in the Amazon drastically reduced deforestation.

Conversely, *destructive* interference occurs when fossil fuel subsidies compete against renewable incentives, when biofuel mandates raise global food prices, or when flat water pricing undermines conservation even as efficiency measures are promoted. Recognizing these interference patterns is essential for effective policy design.

Uncertainty Is Not the Enemy

In quantum systems, uncertainty is a feature, not a flaw. We need provisional wisdom—flexibility, redundancy, and feedback loops—rather than rigid precision. Rather than trying to predict every outcome, we must design solutions that can adapt and thrive under unpredictable conditions.

Designing the Quantum Walk

To design a quantum walk, we must prototype in parallel rather than sequentially, reward learning over fixed outcomes, measure entangled impacts, enable local variations while maintaining global coherence, and invest in uncertain moonshots. This mindset fosters innovation by encouraging multiple experiments, real-time learning, and cross-scale feedback.

Walking Toward the Future

Quantum walks are already shaping sustainability. Indonesia's

Blue Economy Strategy links coastal biodiversity with low-carbon livelihoods. The EU's Circular Economy Laws embed the right to repair and extended producer responsibility. Peru's Indigenous Forest Guardians use blockchain to ensure transparency while supporting local stewardship. The African Union's Great Green Wall initiative regenerates degraded lands while stabilizing migration. Puerto Rico's solar co-ops bring energy sovereignty after disasters.

Mindset Shifts

These examples reflect the necessary shift from ownership to stewardship, precision to resilience, metrics to feedback, policy silos to entangled governance, extraction to regeneration, and control to emergence.

The preceding sections illustrated how quantum thinking reshapes our understanding of uncertainty, experimentation, and systemic interdependence. Yet insight without application risks remaining abstract. To move from theory to transformation, we need a practical bridge—a set of design principles and tools that help leaders, policymakers, innovators, and communities embody quantum logic in daily decision-making. Just as physics translates wave-particle duality into technologies like semiconductors and MRI machines, sustainability must translate entanglement, interference, and observer effects into actionable strategies. The *Quantum Toolbox* serves this purpose: a pragmatic framework for turning complexity into coherence, equipping practitioners with methods to design, evaluate, and scale solutions that are adaptive, systemic, and resilient.

Quantum Toolbox: Principles for Nonlinear Action

The Quantum Toolbox bridges abstract quantum metaphors—such as superposition, entanglement, observer effects, interference, and uncertainty—with practical methods for designing, evaluating, and scaling sustainability solutions. It is meant as a hands-on guide for leaders, policymakers, and innovators who want to translate complex thinking into actionable strategies.

Superposition Framework: Designing Beyond Binary

The first principle is superposition: holding multiple strategies open at once rather than collapsing prematurely into a single "winner." In sustainability, this means exploring diverse solutions in parallel until the system reveals what works best. It requires resisting "winner-takes-all" narratives and ensuring funding mechanisms allow parallel experimentation. Trade-offs can often be redefined as co-benefits, as when urban density and green cover are integrated into holistic design. A practical tool here is the *Superposition Portfolio Map*, a 2 x 2 matrix that plots initiatives based on readiness and systemic leverage, ensuring that at least one high-risk, high-impact initiative remains alive. Germany's *Energiewende*, for instance, deliberately invested in solar, wind, grid technology, and energy storage simultaneously, avoiding premature technological lock-in.

Entanglement Map: Seeing Systems of Systems

Sustainability challenges do not exist in isolation; energy, water, food, finance, and culture are deeply interwoven*. Mapping these interdependencies allows us to see how interventions in one area ripple across others. Begin by identifying three to five domains influenced by your initiative—for example, urban tree planting affects biodiversity, stormwater management, public health, and social equity. From there, draw positive and negative feedback loops and identify hidden leverage points where a small adjustment could unlock multiple benefits[65]. A useful tool is the *Entanglement Diagram* (or *Star Map*), which places the intervention at the center (e.g., EV policy) and maps ripple effects across supply chains (rare earth mining and labor conditions), energy systems (grid carbon intensity), social systems (public transit vs. private vehicle ownership), and the economy (jobs, equity, and investments).

Observer Effect Audit

Metrics are not neutral; they shape behavior and system incentives. The observer effect principle reminds us that measuring the wrong thing can distort priorities. When designing metrics, ask what

*This set of categories highlights sectoral interdependence—particularly the interplay of finance and culture alongside resource flows. Within SCI, these domains are captured under planetary boundaries, social equity, and meaning/engagement. Appendix XVII Table XVII.2 illustrates this alignment – e.g. how finance and culture map onto innovation health and meaning/engagement in SCI.

behaviors the metric encourages, whether it prioritizes short-term gains over long-term resilience, and what important dimensions remain invisible—such as cultural values, biodiversity, or equity. For example, instead of tracking only the number of EVs sold, consider measuring net reductions in transportation emissions per capita to capture a truer picture of systemic impact.

Interference Pattern Analysis

Policies, like waves, can interfere with one another—either amplifying or canceling effects. A constructive interference occurs when overlapping policies, such as feed-in tariffs combined with job creation mandates, accelerate renewable adoption while boosting local economies. Conversely, destructive interference arises when fossil fuel subsidies undermine renewable incentives, or when biofuel mandates unintentionally raise global food prices. The *Policy Wave Map* helps identify such interactions by visualizing overlapping interventions and adjusting their sequencing or bundling to maximize synergy while minimizing conflict.

Uncertainty Design Lens

In quantum thinking, uncertainty is not a flaw but an input to be leveraged. Effective strategies build flexibility, redundancy, and real-time feedback loops rather than seeking rigid precision. Scenario exploration—developing three to five plausible futures instead of a single forecast—helps prepare for uncertainty. Stress-testing asks "What if this solution fails under unexpected conditions?" and designing redundancy ensures we do not optimize everything at the expense of resilience. Singapore's water security strategy is a prime example, blending desalination, rainwater harvesting, recycling, and demand reduction to hedge against unpredictable climate futures.

Quantum Walk Canvas (1-Page Framework)

The *Quantum Walk Canvas* is a simple yet powerful design tool that integrates all the principles above. It includes mapping stakeholders, flows, and feedbacks; listing three to five simultaneous strategies (superposition paths); identifying cross-domain impacts (entanglement web); auditing metrics for observer effect risks; mapping synergies and conflicts (interference map); and exploring uncertainty levers through scenarios and adaptive pathways. This one-page canvas is ideal for workshops or strategy sessions to ensure that decision-making remains nonlinear, systemic, and iterative.

Quantum Questions for Leaders

Before launching any sustainability initiative, leaders should challenge themselves with a set of guiding questions: What unseen connections might amplify or undermine this effort? If we delay deciding, what new possibilities might emerge? How might our chosen metrics distort what we truly value? Could multiple small interventions outperform one large, centralized solution? And finally, what would this look like if we designed it not for control but for emergence, allowing systems to adapt and co-create solutions?

Transition to Application

The next part of this book explores how quantum thinking transforms practical solutions—from bio-based materials to digital governance, and from finance to food systems.

Applying the Quantum Walk Canvas to Urban Water Security

Stakeholders, Flows, and Feedbacks

Municipal water utilities, households, industries, farmers, NGOs, and climate agencies interact across a shared basin. Feedback loops include droughts triggering conservation campaigns, price hikes affecting equity, and floods overwhelming existing infrastructure.

Superposition Paths

- Expand rainwater harvesting at household and community levels.
- Invest in decentralized wastewater recycling for industrial clusters.
- Restore upstream wetlands as natural buffers.
- Pilot dynamic pricing for high-volume users while protecting basic human rights to water.

Entanglement Web (Cross-Domain Impacts)

- **Energy** – Desalination or pumping groundwater increases electricity demand.
- **Food** – Agricultural irrigation depends on shared water.
- **Equity** – Rising costs risk burdening marginalized groups.
- **Culture** – Community rituals and Indigenous stewardship shape trust and compliance.

Observer Effect Audit

Current metrics emphasize "liters supplied per capita," which incentivizes supply expansion rather than conservation. Shifting to "liters saved per capita" or "ecosystem services restored" realigns incentives toward resilience.

Interference Map (Synergies and Conflicts)

- **Constructive interference** – Pairing wetland restoration with flood insurance schemes amplifies resilience.
- **Destructive interference** – Subsidizing groundwater pumping undermines investments in water reuse.

Uncertainty Levers

Scenario planning considers:

- prolonged drought (stress test)
- sudden infrastructure collapse from extreme flooding
- rapid tech adoption (low-cost recycling or atmospheric water harvesting)

Strategies emphasize redundancy and adaptive governance to stay resilient under multiple futures.

Chapter 5

From Leverage Points to Entanglement

Reframing Intervention in a Nonlinear World

> The trick, it turns out, isn't to push harder on the lever. It's to understand where the fulcrum lies, and how the entire machine bends with it.

The Legacy of Donella Meadows

In 1999, systems thinker Donella Meadows published her landmark essay *Leverage Points: Places to Intervene in a System*[66], where she identified 12 leverage points ranging from shallow — least powerful (adjusting parameters) to deep —most powerful (shifting paradigms)[67].

1. **Constants, parameters, numbers** (e.g., subsidies, taxes, standards).
2. **Sizes of buffers** (relative to their flows).
3. **Structure of material stocks and flows.**
4. **Lengths of delays.**
5. **Strength of negative feedback loops, relative to the impacts they are trying to correct against.**
6. **Gain around driving positive feedback loops.**
7. **Structure of information flows.**
8. **Rules of the system** (e.g., incentives, punishments).
9. **Power to add, change, or self-organize system structure.**
10. **Goals of the system.**
11. **Mindset or paradigm** out of which the system arises.
12. **Power to transcend paradigms.**

Her critical insight was that the most transformative interventions happen at the level of goals, mindsets, and paradigms rather than simply tweaking rules or numbers*.

Meadows' hierarchy has profoundly shaped sustainability practice. Yet 25 years later, with climate and ecological crises intensifying into existential threats, the framework must evolve. Leverage alone is no longer enough; we must also account for how interventions ripple across entangled and interdependent systems.

Classical Leverage Meets Complex Reality

In traditional systems, leverage points function like linear dials—raising the carbon tax rate adjusts a parameter, expanding battery storage strengthens a buffer, and shortening policy feedback cycles reduces delays. Yet complex adaptive systems (CAS)[68] do not behave like classical machines. Every dial is interconnected with others, often in subtle or hidden ways. Feedbacks within these systems are dynamic and self-modifying, and a policy aimed at one sector may unintentionally create disruptions in another. In CAS, interventions may be amplified, muted, or even resisted in nonlinear and unpredictable ways. To be effective, we must move beyond the idea of simply identifying where to act and instead examine how actions interact, interfere, and entangle across entire networks of relationships.

Toward a New Grammar: Entangled Interventions[69]

Entanglement highlights that elements of a system are not independent but interdependent and correlated across space, time, and scale. In complex adaptive systems, no action exists in isolation. Every

* This framework is an earlier version of SCI itself, with nearly identical categories. The SCI measure presented later and further elaborated in Appendix XVII Table XVII.2, builds upon and refines this foundation, ensuring continuity across the book.

intervention resonates across domains, creating ripples, feedback loops, and often unintended consequences. Entanglement means that elements of a system are correlated across scales—economic, ecological, cultural, and technological—even when they appear separate.

Take carbon pricing as an example. On the surface, it is a parameter adjustment—raise the price per ton of CO_2 and let markets react. But in reality, it also shifts energy investments (altering infrastructure flows), consumer behaviors (feedback loops), and social narratives of responsibility (paradigms and mindsets). It interacts with trade policies, equity concerns, and even international geopolitics. In short, a price signal is not just an economic tool—it is an entangled signal within a multidimensional system.

Similarly, clean manufacturing mandates may accelerate green innovation but simultaneously pressure small enterprises, reshape global supply chains, alter labor markets, and change energy demand curves in ways that ripple through entire economies. These mandates interact with global trade frameworks (e.g., carbon border adjustments), which in turn influence climate diplomacy and competitiveness narratives.

Real-world entangled interventions offer even richer lessons. Regenerative agriculture and biochar application not only sequester carbon but also enhance soil fertility, reduce input costs, and build local resilience. CRISPR-edited low-lignin trees[70] lower the chemical and energy footprint of pulping while reducing deforestation pressure, enabling carbon-negative materials for packaging and textiles, and fostering rural economic opportunities. Rainwater harvesting and urban cooling initiatives—such as rooftop gardens or permeable pavements—simultaneously reduce flood risk, decrease water demand, and mitigate urban heat islands while rekindling cultural practices tied to community water stewardship. Smart EV charging (vehicle-to-grid or V2G systems) stabilizes grids, creates income streams for drivers, and improves job access, thereby intertwining mobility, equity, and energy transitions. Climate storytelling, when rooted in local narratives, builds policy legitimacy, influences funding priorities, and embeds cultural meaning into decision-making processes. Degrowth movements and cooperative economies reduce material throughput while restoring ecological balance and redefining prosperity as sufficiency rather than accumulation. Satellite monitoring combined with Indigenous land maps protects ancestral territories, amplifies traditional stewardship, and unlocks climate finance, showing how modern technology can complement and elevate ancient wisdom[71].

This entangled thinking extends to finance and governance. Climate-linked fiscal transfers, where states receive funding based on ecological performance, encourage decentralized innovation and align economic priorities with planetary health. Central banks incorporating climate risk into monetary policy redirect capital away from high-carbon sectors, reshaping financial flows and reinforcing systemic resilience. Polycentric water governance, in which village-level management nests within larger catchment systems, reduces conflicts and fosters sustainable extraction. Participatory digital twin models for urban planning allow communities to co-create future scenarios, enhancing both technical and social resilience through transparency and iterative design.

These examples reveal that leverage is not a single point but a web of relationships. Effective interventions don't just "fix" one part of the system; they activate multiple feedbacks, alter mindsets, and shift system goals in parallel. Entangled interventions cut across several of Meadows' leverage points[72] at once—adjusting rules, influencing information flows, reshaping system goals, and sometimes even challenging the underlying paradigm itself.

The Entangled Leverage Framework

Meadows' original framework can be extended into a two-dimensional matrix that considers both the depth of an intervention (from shallow parameters to deep paradigm shifts) and its entanglement potential (from localized to systemic). A shallow, low-entanglement intervention might be a simple carbon tax rate adjustment, while systemic and deeply entangled levers include global fossil fuel divestment campaigns or regenerative well-being economies that reshape values, policies, and infrastructures simultaneously. Building codes, circular economy mandates, and border carbon adjustments occupy the middle layers. The deepest, most entangled levers are the most transformative—but also the most unpredictable.

Designing with Entangled Leverage

Designing interventions for entangled systems requires a new mindset. Indirect leverage must be recognized: women's education not only reduces fertility rates but enhances climate resilience, fosters economic innovation, and accelerates cultural transformation. Similarly, cultural narratives can shift paradigms faster than legislation. Cross-leverage synergies should be cultivated by combining digital carbon wallets with dynamic pricing and sustainability education or pairing bioeconomy investments with waste valorization. Avoiding policy echo chambers is equally critical—policies must create enabling constraints that encourage self-organization and local adaptation instead of rigid, top-down control.

Reflexivity: Observing the Observer

Entangled leverage implies reflexivity: the system not only responds to the intervention but also to the intervenor. In such systems, the act of observation itself influences outcomes. Climate models, for instance, are not neutral—they embed assumptions about feedback sensitivity, economic behavior, and political will. The way we model climate futures often shapes which policies are considered viable or politically acceptable. Similarly, national or corporate net-zero roadmaps frequently rest on the expectation that negative emissions technologies, such as direct air capture (DAC), will be available at scale in the future. Here, belief in technology becomes part of the trajectory, with imagined solutions influencing present strategies and investment flows.

ESG ratings offer another vivid example of reflexivity. The criteria selected by rating agencies reflect particular value systems—such as prioritizing carbon reduction over biodiversity preservation or labor rights—and these choices guide investor behavior, corporate disclosure strategies, and market narratives. Likewise, when cities test bike-sharing programs through pilot projects, the act of measuring "success" can itself alter public engagement, media narratives, and eventual infrastructure priorities. Observation does not merely record reality; it reshapes it.

This dynamic also appears in academia, where metrics like journal impact factors or citation counts are used as proxies for research quality. Such measures influence what researchers choose to study, often nudging innovation toward what is measurable or trendy rather than what is transformative. Similarly, carbon accounting standards wield power by defining what counts as "Scope 3 emissions" or what qualifies as "renewable," thereby shaping the behaviors of firms, governments, and entire industries based on the framing of the observer.

Even at the grassroots level, NGO monitoring and evaluation can create reflexivity. In community-led conservation projects, local actors may adjust their behavior simply because they know they are being evaluated. The mere presence of observers and metrics changes the dynamics of participation, cooperation, and perceived success.

This is the quantum observer effect in systems change: our intentions, language, and framing do not just measure systems—they actively co-create the responses we observe.

Tools for Entangled Thinking

Our old tools are often too linear for complex realities. KPIs give way to multi-dimensional dashboards; cost-benefit analysis is replaced by cross-impact balance analysis; linear planning shifts to adaptive pathways and scenario trees; risk analysis evolves into resilience and option-value frameworks; and sectoral models expand into agent-based, multi-nodal simulations. These tools embrace complexity rather than oversimplifying it.

Case Study: The eCarbon Card[73] as an Entangled Lever

My own work with the *eCarbon Card*, offers a vivid example. It is not just a digital wallet—it is:

- behavior modifier (nudges)
- a data platform (transparency)
- a governance tool (accountability)
- a signal for equity (tiered allowances)
- a potential market instrument (carbon trading)

Its impact simultaneously spans sectors: transport, energy, retail, data privacy, digital literacy, social equity, and political trust, demonstrating how a single intervention can reverberate across sectors and feedback loops.

It embodies entangled leverage—a deep, cross-sectoral, paradigm-shifting intervention designed for emergence.

Transcending Paradigms

"There is no single 'paradigm of sustainability.' The most powerful leverage lies in our capacity to see that we are inside a story—and choose to tell a new one."

At the highest level of Meadows' framework lies paradigm shift—the ability to see the system through a fundamentally different lens. This means moving beyond carbon as the sole metric to embrace planetary well-being, prioritizing sufficiency and regeneration over growth, and integrating plural knowledge systems, including Indigenous, feminist, and ecological worldviews. It also involves rewriting cultural narratives from extraction to reciprocity and valuing slowness, attention, and interdependence in both policy and design. These and other examples are detailed below:

- **Moving beyond carbon as the only metric** – Broaden focus to planetary health indicators such as biodiversity, nutrient cycles, and cultural vitality.
- **Shifting from scarcity-driven economics to sufficiency-driven living** – Redesign consumption models to prioritize enough rather than more, emphasizing quality of life over GDP.

- **Reframing success from control to co-evolution** – Embrace adaptive governance, iterative experimentation, and learning networks instead of rigid planning.
- **Recognizing knowledge as plural** – Value Indigenous, feminist, and ecological epistemologies alongside scientific and economic ones.
- **From "saving the planet" to "healing our relationship with it"** – Reconnect ethics, awe, humility, and stewardship to how we design interventions.
- **Letting go of fixed roles and identities** – Encourage scientists to be storytellers, policymakers to be listeners, and communities to be designers.
- **Rewriting cultural narratives** – Replace extraction and conquest with restoration, reciprocity, and intergenerational responsibility.
- **Valuing slowness and attention** – Challenge the acceleration logic of markets and tech in favor of slow growth, local time, and care.

Preparing for the Walk Ahead

To navigate uncertainty with coherence, we must deepen systems literacy, cultivate pattern recognition, and practice strategic humility. Iteration becomes a core discipline, and emotional resilience is essential for transdisciplinary teams to thrive. Leadership in this space is not about dominance but about creating space for emergence. Awe and joy, often overlooked, are powerful catalysts for sustained engagement and creativity. These and other examples are detailed below:

- **Deepen your systems literacy** – Learn to map loops, sense thresholds, and think in delays—not just in diagrams, but in daily decisions.
- **Practice strategic humility** – Acknowledge the limits of foresight and embrace iteration, feedback, and course correction.

- **Embody the future now** – Live as though the regenerative, just world you imagine is already possible—design from the future backward.
- **Develop your "pattern sensing" capacity** – Pay attention to emergence: What is bubbling up? What is dying away? What is connecting?
- **Build emotional and relational resilience** – Engage in practices (e.g., council, listening circles, systems constellations) that strengthen community bonds and meaning.
- **Join or create transdisciplinary teams** – Bridge silos—merge art and science, policy and practice, intuition and data.
- **Be ready to fail forward** – Treat breakdowns as information. Design with buffers and redundancy.
- **Stay grounded in place and people** – Walk the land, listen to those affected, design from the bottom-up—not the abstract center.
- **Learn to lead from the middle** – Influence across boundaries, seed conversations, convene rather than command.
- **Remember joy, awe, and gratitude** – They are not distractions from the work—they are the fuel that sustains the long journey.

Vignettes of Entangled Leverage

Women-led solar microgrids in Bihar have transformed not just energy access but local economies and social roles. Regenerative agriculture in Kenya has improved soil health while altering cultural narratives and economic feedback loops. Bus Rapid Transit in Bogotá reduced emissions while enhancing equity, safety, and urban identity. Mycelium bio-design initiatives in New York fuse material innovation with Indigenous cultural renewal. Finland's climate curriculum integrates science and mental health to cultivate agency rather than apathy.

Entangled interventions are not about finding a single fulcrum but understanding and activating the web of relationships that define our interconnected world.

Chapter 6

Systemic Entanglement

Food, Water, Energy, Carbon, Equity—The New Physics of Sustainability

> In quantum systems, nothing exists in isolation. In sustainable systems, that truth becomes moral, not just mathematical.

The False Promise of Silos

Modern development strategies often divide the world into discrete domains—water policy, energy infrastructure, food security, carbon reduction, and poverty alleviation—each with its own budget, ministry, and metrics, as though they were separate crises. Yet reality refuses such partitions. The food system cannot function without water; energy transitions are inseparable from equity and carbon; poverty alleviation is entangled with climate adaptation and resource access. Just as quantum particles exhibit nonlocal correlations, sustainability challenges are causally linked across scales, sectors, and values. Every intervention in one domain ripples through others, often in ways linear thinking fails to anticipate. To create resilient strategies, we must learn to map and work with these entanglements rather than against them.

The Entangled Pentad: Food–Water–Energy–Carbon–Equity

The future of sustainability is defined by five co-evolving systems—food, water, energy, carbon, and equity—each inseparable from the others. Agriculture, for instance, consumes around 70%

of global freshwater withdrawals and contributes roughly 25–30% of greenhouse gas emissions through deforestation, methane from livestock and rice paddies, and nitrous oxide from fertilizers. Food security is tightly connected to energy (fuel for irrigation, transport, and fertilizers) and water (droughts, salinity, floods), while land-use decisions such as allocating cropland for biofuels or carbon sinks often involve equity tradeoffs that impact smallholder farmers.

Water is equally interconnected, playing a vital role in energy production from power plant cooling to hydropower and sanitation, while climate change disrupts rainfall patterns, snowmelt timing, and aquifer replenishment. Energy is needed for water extraction, treatment, and desalination, making water security directly dependent on energy systems. Droughts and water scarcity disproportionately affect marginalized communities, deepening inequality.

Energy systems, still largely reliant on fossil fuels, drive our food and water systems by powering tractors, fertilizer production, and groundwater pumps. Transitioning to renewables affects land availability, mineral extraction, and water demand—such as the cooling requirements of concentrated solar power plants. Electrification can reduce emissions but risks widening energy poverty if access to reliable grids remains unequal.

Carbon serves as a proxy for the metabolism of modern civilization. Every ton of CO_2 tells a story of extraction, transformation, and transport—container ships moving goods across oceans, steel and cement factories running non-stop, forests cleared for commodity crops, and high-speed data centers fueling the digital economy. Decarbonization is therefore not merely technical; it is metabolic. It requires redesigning industrial metabolism (steel, cement, plastics), biological metabolism (agriculture, soil carbon, forests), and urban metabolism (energy, mobility, waste cycles). Reducing emissions means reconfiguring these flows and rethinking the underlying priorities of modern life.

Equity asks the essential questions of who wins, who loses, who pays, and who decides. Solar subsidies may benefit homeowners but exclude renters. EV mandates might price out low-income drivers unless carefully targeted. Carbon taxes without rebates tend to burden the poor. Policies designed without Indigenous voices risk violating cultural and land rights. Sustainability is never neutral; it must be both green and fair.

Cascading Entanglements[74]: Case Examples

The complexity of these entanglements is best illustrated through real-world examples.

Biofuels and Food Prices

Biofuels, promoted to reduce fossil fuel dependence, often divert land from food to fuel crops such as corn and sugarcane. This can raise food prices, place additional pressure on already stressed aquifers due to irrigation demands, and trigger deforestation in biodiverse regions, resulting in net emissions gains. The 2008 food price crisis demonstrated how energy policy can destabilize both food security and equity.

Groundwater Depletion in Punjab, India

In Punjab, India, subsidized electricity incentivizes over-pumping of groundwater. As aquifers deplete, energy demand and emissions rise while poor farmers—lacking deep borewells—lose access first, worsening inequality. Degraded aquifers also weaken resilience to droughts and floods. Potential solutions, such as solar pumps or drip irrigation, require coordinated alignment across water, energy, and social policies.

Reforestation and Water Availability

Reforestation for carbon offsets can reduce local groundwater availability in semi-arid regions when non-native, fast-growing trees are planted. These trees alter albedo and evapotranspiration, shift microclimates, and create land tenure conflicts if local communities are excluded from planning. A carbon win can easily become a hydrological or social loss without systems-aware design.

Urban Bike Lanes and Gentrification

Urban bike lanes, while reducing emissions and improving quality of life, can also trigger gentrification. In some cities, property speculation and rising rents displace marginalized residents, depriving them of the benefits intended for all.

Subsidizing EVs Without Grid Reform

Similarly, EV subsidies without grid reform can shift emissions upstream if electricity still comes from fossil fuels, benefiting wealthier consumers first and reinforcing energy inequality. Without plans for battery recycling or second-life applications, mining pressures and environmental impacts persist.

Constructive Entanglement: Agroecology

Conversely, constructive entanglements can yield synergistic benefits. Agroecological farming systems restore soil carbon, increase water retention, enhance biodiversity, and diversify rural incomes, while local seed-saving initiatives strengthen cultural resilience.

Constructive Entanglement: Free Public Transit in Tallinn

Free public transit in Tallinn reduced emissions, improved social equity, boosted local economic activity, and enhanced civic trust, proving that equitable policies can deliver systemic gains.

Constructive Entanglement: Soil Carbon Practices as a Multiplier

Soil carbon practices such as cover crops, reduced tillage, and agroforestry create feedback loops that stabilize yields, improve water retention, and generate carbon credits for farmers, accelerating adoption as farms become more resilient.

Superposition in Strategy: Designing for Multiple Outcomes

Strategic intervention or design can generate simultaneous benefits across the five entangled domains – Carbon, Water, Energy, Food and Equity. Agroforestry corridors, for example, reduce carbon emissions, increase water retention, lower agricultural input needs, raise crop yields, and improve rural livelihoods. Solar microgrids cut carbon emissions while enhancing energy access, enabling cold storage, and strengthening local ownership. Regenerative agriculture boosts soil water capacity, reduces fertilizer use, improves nutrition, and restores land health while generating carbon reductions. Perovskite solar technologies promise affordable renewable energy with low lifecycle emissions, while community-owned EV charging builds energy resilience, reduces local carbon emissions, and creates jobs. Similarly, bioremediation of brownfields reduces toxic pollution, restores aquifer health, improves land safety, and advances urban environmental justice.

From Tradeoffs to Synergies

Sustainability challenges have long been framed as tradeoffs—clean energy versus cheap energy, growth versus emissions, forests versus farming, wind power versus bird safety, or mineral mining versus Indigenous rights. Quantum design allows us to replace either/or choices with both/and solutions. Agrovoltaics combine crops and solar energy on the same land, while smart wind siting uses radar-guided turbine pauses to protect migratory birds. Nature-based urbanism with vertical gardens, green roofs, and micro-parks fosters biodiversity while supporting high-density living. Circular textile hubs create local jobs, reduce waste, and cut reliance on virgin materials.

Stacking functions across domains further amplifies impact. Mangrove restoration provides coastal defense, supports fisheries, and acts as a powerful carbon sink. Solar irrigation cooperatives deliver water security, clean energy, and rural employment in one intervention. High-leverage structures integrate planetary boundaries into national budgets and tie urban zoning incentives to biodiversity

* These operational categories are best understood as measurable expressions of SCI's planetary boundaries (carbon, water, energy, food) and social equity dimensions. Appendix XVII Table XVII.2 shows this mapping.

corridors. Feedback-sensitive measures like dynamic congestion pricing based on air quality or digital twins for grid balancing during climate stress events ensure responsiveness. Mindset shifts, such as citizen climate assemblies or "carbon diets" that frame low-emission living as empowerment, reshape culture and policy together.

Governance for Entangled Domains

Effective governance must evolve from departments to interconnected nodes and from silos to interoperable platforms. Metrics should give way to mapping tools that capture cascading impacts. Planning must shift to sensing, using real-time data feedback loops to respond dynamically. Power itself must become participatory, with inclusive and co-designed policies ensuring diverse voices and perspectives shape decisions.

Ethical Implications of Entanglement

Entanglement is not merely technical but also moral. Your EV battery may depend on cobalt mined under exploitative conditions. Your carbon offsets may displace Indigenous communities. Your textile choices can drive water scarcity in dyeing hubs across Asia. True sustainability must be rooted in responsibility, empathy, and visibility across supply chains and social systems.

Walking Forward Together

In navigating complexity, we are not powerless. Leadership in this context is less about issuing commands and more about choreographing collective action. This requires embracing iteration over perfection, fostering collective feedback and learning, and designing with humility, joy, and pluralism as guiding principles.

What Comes Next

The next part of this book explores quantum-aligned solutions that reflect these entangled realities, ranging from digital interventions like the eCarbon Card[75] to biological innovations such as CRISPR-edited low-lignin trees[76] and mycelium-based packaging[77, 78], as well as financial and governance tools designed for systemic resilience and transformation.

Part III: Solutions Beyond the Obvious

Chapter 7

From Linear Solutions to Quantum Interventions

Designing for Complexity

Complex problems are rarely solved by single-point solutions. They are transformed through patterns, relationships, and emergent strategies—what I call quantum interventions.

The Illusion of Linear Solutions

For decades, our sustainability approach has been framed through the lens of reductionist problem-solving: identify a single issue, design a solution, scale it. This linear mindset emerged from the industrial and Newtonian worldview, where systems were assumed to respond predictably to inputs, much like machines. It gave us impressive breakthroughs—vaccines, water treatment, mass production—but it also created the very global crises we face today. Climate change, biodiversity loss, and inequality are not isolated problems that can be "fixed" with one innovation or policy lever. They are manifestations of deeply entangled, adaptive systems.

Linear thinking continues to dominate much of the sustainability discourse, where carbon capture is often seen as the silver bullet, electric vehicles are heralded as the ultimate solution, and net-zero pledges are treated as the final metric of success. Each of these interventions is undeniably valuable, yet they remain insufficient when pursued in isolation. Without systemic context and integrated design, such measures risk shifting problems rather than solving them—creating rebound effects and unintended consequences that undermine long-term progress. A carbon capture facility, for instance, may successfully remove CO_2 from the atmosphere, but if it draws power from coal, it simultaneously increases local air pollution and entrenches fossil-based energy systems. Similarly, a city that mandates widespread EV adoption without enhancing public transit or walkable infrastructure may reduce tailpipe emissions while deepening mobility inequities, increasing urban congestion, and perpetuating car dependency.

Quantum interventions recognize that complex systems cannot be effectively addressed with single-threaded, linear solutions. They demand strategies that are inherently relational—built on the understanding that changes in one domain ripple across others—adaptive, responding to feedback and shifting conditions rather than rigid plans, and multi-dimensional, targeting interconnected challenges instead of single issues in isolation. The purpose of this chapter is to move beyond the *illusion* of linearity and explore how to design interventions that not only acknowledge these entangled realities but thrive within them, leveraging feedback, co-benefits, and emergent opportunities to create lasting systemic transformation.

What Makes an Intervention 'Quantum'?

A quantum intervention is one that operates across multiple domains simultaneously, leveraging systemic feedbacks and emergent properties. Instead of seeking optimization for one metric, such as cost per ton of carbon reduced, it pursues patterns of benefits that unfold across interconnected systems—social, ecological, technological, and economic.

The essence of a quantum intervention lies in its ability to hold **superposition**—the capacity to address multiple possibilities and pathways at once—while also being **entangled** with other systems and **sensitive to observation and feedback**. A well-designed intervention is not a rigid solution but a living experiment that adapts as the system responds.

Consider regenerative agriculture. It is not just a way to grow food; it sequesters carbon, improves water retention, boosts biodiversity, enhances farmer livelihoods, and strengthens community resilience. Its impact emerges from the interplay of biological, social, and economic processes. Similarly, distributed energy systems like solar microgrids are not only about electricity generation. They create local ownership models, reduce emissions, improve energy sovereignty, and catalyze economic development. These are quintessential quantum interventions—small in appearance, but powerful in their ability to shift entire patterns of interaction.

Characteristics of Quantum Interventions

A quantum intervention is distinguished by several attributes:

Multi-domain entanglement

A single action creates positive ripple effects across multiple systems. For example, planting urban forests simultaneously mitigates heat islands, improves air quality, increases biodiversity, and boosts mental health for residents.

Constructive interference

When multiple actions are aligned, they amplify each other instead of competing. An example is pairing EV adoption with grid decarbonization and public charging infrastructure, creating systemic resilience rather than technological silos.

Observer responsiveness

The way interventions are framed, measured, or narrated alters how people engage with them. This reflects the "observer effect" at the societal scale—metrics shape behavior, and behavior shapes outcomes.

Emergent adaptability

Quantum interventions are designed for iteration and learning, not for static implementation. A city's climate resilience strategy, for instance, may involve pilot zones or living labs where adaptive governance is prioritized over prescriptive plans.

From Pilots to Patterns

Much of the sustainability world is obsessed with "pilots" and "proof of concept," but too often these pilots remain isolated, unable to scale or evolve. Quantum interventions, in contrast, focus less on a singular breakthrough and more on establishing patterns of systemic change that can replicate across contexts.

Consider the example of community-owned microgrids in Puerto Rico. After Hurricane Maria, local cooperatives and NGOs helped install solar-powered microgrids that provided reliable electricity when centralized infrastructure failed. What began as an emergency response evolved into a model of energy sovereignty, blending technical resilience with social equity. These microgrids are now inspiring similar initiatives in other disaster-prone regions, illustrating how a local intervention can scale through patterns, not just replication.

Another example comes from the Netherlands' "Room for the River" program. Instead of building higher dikes to control floods—a linear, reactive solution—Dutch planners redesigned entire river landscapes to give water more space to flow naturally. This intervention not only reduced flood risks but created new public spaces, revitalized ecosystems, and improved urban quality of life. It is a living example of how systemic redesign can yield multiple co-benefits beyond the original problem statement.

Superposition in Policy Design

Quantum interventions demand that policymakers resist the urge to collapse choices prematurely. Instead of choosing between renewable subsidies or carbon pricing, or between hydrogen and electrification, a superposition mindset holds multiple options open until systemic feedback reveals which pathways are most effective.

For instance, Germany's Energiewende combined feed-in tariffs with local manufacturing incentives, social acceptance programs, and grid modernization. Rather than betting on a single technology, it cultivated a portfolio of renewables—wind, solar, biomass—and allowed public participation to guide adoption. Similarly, Singapore's green urban design strategy integrates vertical gardens, energy-efficient architecture, and water-sensitive infrastructure, creating an urban ecosystem where multiple solutions coexist and reinforce one another.

Case Studies of Quantum Interventions

Mycelium-Based Materials and Circular Economies[79, 80]

Mycelium packaging solutions, made from agricultural or paper mill waste, replace fossil-based foams while reducing landfill waste. When developed as part of a local circular economy, they create jobs, valorize industrial by-products, and align with nature-based design principles. This is not simply material substitution but a redesign of economic and biological flows—turning waste into resource and connecting industries like agriculture, paper, and packaging in regenerative loops.

eCarbon Card[81]: A Behavioral and Market Tool

The eCarbon Card, a digital carbon wallet, is more than a carbon accounting device. It nudges individuals towards low-carbon behaviors, creates transparent data flows for policymakers, and can integrate with carbon markets to create equitable credit trading systems. Its strength lies in its entanglement with energy use, mobility, retail, and public trust—making it a deep lever for both personal and systemic transformation.

Agrovoltaics and Water Management

Combining solar panels with agriculture creates energy and food security simultaneously. Panels provide shade that reduces water evaporation, while the energy generated can power irrigation or cold storage. When coupled with local ownership and biodiversity corridors, agrovoltaics become a multifunctional system that aligns with community resilience and carbon mitigation.

Biochar for Soil Health and Carbon Sequestration

Biochar, a stable form of carbon produced from pyrolyzed biomass, is applied to agricultural soils to enhance fertility, retain moisture, and lock away carbon for centuries. When integrated into regenerative farming systems, biochar addresses multiple challenges at once—improving crop yields, reducing the need for synthetic fertilizers, and sequestering atmospheric CO_2 in a stable form. Local biochar production using agricultural residues closes material loops, reduces waste burning, and creates rural livelihood opportunities, making it both a carbon sink and a socio-economic catalyst.

Seawater-Based Desalination Powered by Offshore Wind

Coupling offshore wind farms with advanced, energy-efficient desalination plants transforms two critical resources—clean energy and fresh water—into a single integrated system. The wind power drives reverse osmosis or forward osmosis processes, producing potable water for coastal communities while avoiding fossil-fuel-based energy use. Waste brine is managed through ecological dispersal or valorized for mineral recovery, creating new revenue streams. This multi-sector synergy reduces water stress, enhances energy security, and decarbonizes both sectors simultaneously, embodying a deeply entangled intervention in climate adaptation and mitigation.

Designing Quantum Interventions

To design for quantum leverage, we need a shift from single metrics (e.g., cost per unit of energy) to multi-dimensional frameworks that assess co-benefits, equity impacts, and long-term resilience. This design process involves:

- **Mapping entanglements** – Understand how energy, water, food, carbon, and equity interact in a given context.
- **Building feedback loops** – Integrate real-time sensing, monitoring, and adaptation into every intervention.
- **Engaging diverse knowledge systems** – Combine scientific data with local, Indigenous, and community narratives to ensure legitimacy and cultural fit.

- **Designing for co-benefits** – Seek interventions that improve multiple aspects of the system simultaneously—ecological health, social well-being, and economic vitality.

For instance, regenerative agriculture pilots should not only track yields or soil carbon but also monitor groundwater levels, farmer incomes, and community health metrics. Similarly, urban climate strategies should integrate housing affordability, public health, and biodiversity alongside emissions reduction.

The Role of Narrative and Culture

No intervention can succeed without a cultural narrative that resonates. Quantum interventions rely on stories, metaphors, and shared meanings to align stakeholders across sectors and scales. A project like Colombia's TransMilenio BRT system was not just about buses—it was about reclaiming public space, reducing inequality, and reshaping the narrative of urban life. When combined with public art, participatory planning, and community education, its impact extended far beyond transportation.

Similarly, Indigenous fire management practices in Australia are now recognized as both cultural and ecological interventions, preventing megafires while restoring sacred landscapes. These approaches remind us that interventions are never just technical—they are emotional, symbolic, and cultural.

Traditional Ecological Knowledge: Quantum Wisdom in Practice

Climate action is not only a question of advanced technologies or new governance models; it is also about recovering and valuing ways of knowing that have sustained ecosystems for millennia. Indigenous knowledge systems, often grouped under the term Traditional Ecological Knowledge (TEK), embody principles of nonlinearity, resilience, and entanglement long before these terms entered the vocabulary of science. Where linear models reduce cause and effect, TEK embraces webs of reciprocity, cycles, and thresholds—what this book calls *quantum wisdom*.

More than Cultural Symbolism

Too often, Indigenous practices are acknowledged in climate policy only as cultural heritage or discussed in terms of avoiding harm to communities. But TEK offers more than symbolic value; it provides **practical strategies for sustainable land stewardship**. For example, Indigenous Australians have long practiced cultural burning, a form of mosaic fire management that reduces catastrophic wildfire risk, regenerates biodiversity, and enhances soil fertility. This practice reflects quantum principles of feedback: small, deliberate interventions prevent large-scale collapse.

Systems of Reciprocity and Feedback

Other examples abound across the world. In New Zealand, the Māori principle of *kaitiakitanga* frames humans as guardians of forests, rivers, and seas—responsibilities that emphasize feedback loops and thresholds in ecological balance. In the Andes, Indigenous terracing and crop-rotation systems distribute water, reduce erosion, and maintain biodiversity across altitude gradients, ensuring resilience against drought and floods. In **North America**, First Nations' fisheries management involved seasonal restrictions, habitat enhancement, and spiritual governance, anticipating the concept of planetary boundaries by embedding ecological limits into social and cultural norms.

TEK as Quantum Practice

Traditional knowledge resonates deeply with the quantum frameworks advanced in this book:

- **Entanglement** – Human and non-human systems are inseparably linked, with responsibilities flowing both ways.
- **Emergence** – Stewardship practices create ecological conditions (soil health, species balance) that in turn support cultural continuity.
- **Observer Responsiveness** – Ritual, ceremony, and narrative serve as monitoring systems, ensuring people remain attuned to subtle environmental signals.
- **Constructive Interference** – When TEK meets scientific

methods—such as integrating satellite data with local fire knowledge—the result can be solutions stronger than either approach alone.

TEK Principles Aligned with SCI

Ecological Integrity

- "*Fire as Medicine*" – Australian Indigenous fire management views controlled-burns not as destruction but as stewardship, preventing catastrophic wildfires and regenerating biodiversity. This echoes SCI's focus on ecological integrity by sustaining landscape health across generations.

Social Equity

- "*Seven Generations Principle*" – Many Indigenous communities (e.g., Haudenosaunee Confederacy) evaluate decisions by their impact on the seventh generation to come. This intergenerational ethic aligns with SCI's social equity lens, ensuring fairness not just across groups today, but across time.

Circularity & Regeneration

- "*Nothing Wasted*" – Pacific Islander fishing traditions limit harvests to ensure reef regeneration and use all parts of a catch. This principle exemplifies SCI's circularity and regeneration category, embedding balance between use and renewal.

Meaning & Engagement

- "*Land as Kin*" – In Māori philosophy (Whakapapa), land, rivers, and mountains are relatives, not resources. This worldview strengthens SCI's "meaning and engagement" category by showing that sustainability isn't just technical but rooted in values and relationships.

Toward a Braided Future

As Robin Wall Kimmerer writes in *Braiding Sweetgrass*[82], TEK is not only about protecting Indigenous lifeways, but about offering humanity **a blueprint for living with reciprocity and humility**. Quantum sustainability requires precisely this: blending advanced science with Indigenous worldviews to design solutions that are both technically effective and ethically grounded. Rather than treating TEK as a "parallel knowledge system," we can see it as an integral component of the quantum toolkit for climate resilience.

Moving from Pilots to Platforms

A key challenge is scaling quantum interventions without stripping them of their contextual intelligence. Too often, successful pilots are over-standardized in attempts to scale, losing the very flexibility that made them work. Instead, interventions should be treated as platforms for learning and adaptation, where local actors modify and evolve solutions based on real-time conditions.

The African Union's Great Green Wall initiative illustrates this principle. While initially conceived as a tree-planting program across the Sahel, it has evolved into a mosaic of locally adapted agroecological, economic, and cultural projects—ranging from water harvesting to women-led cooperatives. It is less a "project" and more a living network of interventions.

Conclusion: Quantum Interventions as a Mindset

Quantum interventions are not defined by their scale but by their **interconnectedness** and **adaptability**. They invite us to think like gardeners rather than engineers—cultivating conditions for resilience, diversity, and emergence rather than forcing single outcomes.

As we move into the next chapters, we will explore how this mindset applies across key sectors—digital governance, bio-based materials, energy transitions, and finance. Each sector offers opportunities to move from linear problem-solving to quantum system-shaping. The challenge ahead is not to find one perfect solution but to design a field of possibilities where multiple futures can emerge, evolve, and thrive.

Chapter 8

Digital Permutations

From Blockchain to Behavioral Nudges

> We used to build digital tools to simulate the world. Now we must build them to heal it.

The Digital Turn in Sustainability

The digital revolution began with data, accelerated through connectivity, and now stands at a precipice: can it help solve the defining crisis of our era—climate destabilization? While climate change is undeniably physical—a crisis of atmospheric gases, warming oceans, and disrupted energy flows—many of its underlying drivers and solutions are informational. We need to know who emits, when, where, and how decisions are made. We need to understand what incentives nudge systemic change—or block it.

Digital infrastructure offers the ability to detect, connect, and direct these invisible flows. It can transform unseen emissions into actionable insights, anonymous behaviors into accountable choices, and static policies into adaptive learning ecosystems. Yet this is not a techno-utopian promise. In a complex, entangled world, digital tools must be designed not to control, but to guide emergence—building quantum pathways of possibility rather than rigidly optimized ones.

From Centralization to Distributed Intelligence

Traditional governance has relied on hierarchical structures: setting targets, enforcing compliance, aggregating data, and penalizing deviations. This classical model treats systems as linear and controllable. However, quantum-era sustainability requires distributed intelligence—networks of actors that share information, co-create solutions, and adapt in real time.

Digital technologies act as the nervous system of such distributed governance. Blockchain and distributed ledger technologies (DLTs) create decentralized trust and immutable records. IoT sensors, satellite imaging, and AI enable real-time monitoring and feedback loops. App-based behavioral platforms interpret micro-decisions, while smart contracts and APIs allow modular, interoperable governance systems. These tools do not replace policy; rather, they amplify its capacity for responsiveness, granularity, and continuous adaptation.

The eCarbon Card[83]: A Case Study in Behavioral Architecture

Imagine an individual carbon wallet—an instrument that allocates a personal annual carbon budget, tracks emissions from transport, food, utilities, and purchases, and provides real-time feedback on carbon intensity. The **eCarbon Card** is designed to enable such capabilities, allowing users to set goals, compare performance with peers, and even trade or offset surplus emissions.

Unlike surveillance-based systems, the eCarbon Card aims to foster informed agency. By making carbon visible and personal, it uses behavioral tools—gamification, positive social nudges, and tiered incentives—to steer users toward low-carbon choices. It also enables policymakers to implement dynamic tariffs, rebates, or allowances with precision, while protecting vulnerable groups through differentiated thresholds. At the market level, it opens the door for micro-scale carbon trading, and at the national level, it creates a bottom-up emissions map that can refine climate targets.

This is digital permutation at its best—transforming small informational shifts into systemic behavioral changes, aligning individual habits with collective climate goals.

Nudges, Not Nags: The Neuroscience of Sustainable Behavior

The climate crisis is as much about human decision-making as it is about atmospheric physics. People rarely act on logic alone; they are guided by habits, social norms, identity cues, and cognitive biases. Digital platforms have the potential to work with these psychological levers, using insights from behavioral science to create subtle, empowering nudges rather than coercive directives.

For instance, making green energy the default option for utility plans has been shown to significantly increase renewable energy uptake without restricting choice. Real-time dashboards displaying household emissions—or framing achievements in relatable terms like "your trip saved four trees"—shift perceptions of normal behavior. Gamification techniques, such as earning points for eco-friendly transport, or commitment devices like pre-pledged low-carbon months with social visibility, create feedback loops that normalize sustainability. When deployed ethically, such nudges can reshape identity and culture, embedding climate-conscious behavior into the fabric of daily life.

Digital Twins and Scenario Simulation

Digital twins—virtual replicas of physical systems—are emerging as one of the most powerful tools for anticipating complexity. Coupled with AI and real-time data streams, they can model emissions, infrastructure performance, and policy outcomes with remarkable granularity. Unlike static forecasts, digital twins allow decision-makers to simulate millions of micro-interactions, revealing unintended consequences before they occur.

In practice:

- **Energy Systems** – They can dynamically adjust carbon pricing based on renewable energy availability or congestion levels, or nudge consumers to shift loads during peaks—preventing blackouts and lowering emissions.
- **Agriculture** – They can optimize irrigation schedules by integrating weather forecasts, satellite imagery, and soil sensors, improving water use while protecting yields.
- **Urban Design** – Cities like Singapore deploy digital twins to model transport flows, heat islands, and EV adoption before implementing new infrastructure.

- **Industry** – Steel and cement facilities use digital twins to simulate emission controls and integrate carbon capture in production lines.
- **Health and Resilience** – Hospitals and municipalities model how urban heat islands affect admissions, planning cooling interventions accordingly.

In essence, digital twins help create a **reflexive infrastructure**: systems that not only react to changes but also learn, adapt, and evolve through feedback loops. They represent a *quantum walk of possible futures*, where emergent patterns guide actionable strategies in real time.

Blockchain and Carbon Markets: From Trust to Truth

Carbon markets have long suffered from opacity, delayed verification, and a lack of public trust, often plagued by issues like double counting or unverifiable credits. Blockchain technology offers a paradigm shift by providing transparent, immutable ledgers of transactions and automated rule enforcement through smart contracts. By tokenizing carbon credits and regenerative activities, blockchain can democratize access to climate finance.

Smallholder farmers can sell verified soil carbon credits, urban residents can earn micro-incentives for low-carbon mobility, and municipalities can track their Scope 1, 2, and 3 emissions with real-time accuracy. These systems enable small but frequent rewards—micro-incentives—that encourage sustainable behavior, from walking and cycling to taking public transit or using shared electric mobility.

Low-carbon mobility, a critical component of urban decarbonization, includes choices such as electric buses powered by renewable energy, e-bike sharing, carpooling, and optimized rail freight. When integrated with micro-incentive systems—cashback for walking, carbon points for biking, or discounts on sustainable services—such programs not only reduce emissions but also enhance urban health and accessibility.

Digital Risks: When Tools Become Traps

Digital systems are not neutral. If designed without care, they can deepen inequities, promote surveillance, or inadvertently disempower communities. Key risks include data colonialism—where corporations or governments control and profit from emissions data—alongside the digital divide that excludes rural or low-income users who lack devices or internet access. Behavioral nudges can be manipulative if not transparent, and algorithmic optimization often prioritizes narrow metrics at the expense of broader well-being.

To avoid these pitfalls, digital climate tools must embrace plurality, context, and reversibility. Systems should be designed to empower rather than extract, ensuring that data ownership, privacy, and equity remain central principles.

Toward Ethical, Entangled Digital Infrastructure

A quantum-inspired digital climate strategy requires not only technical sophistication but also a values architecture. The infrastructure must be entangled—capturing the co-benefits across carbon reduction, equity, and health. It must be participatory, co-designed with the communities it serves. It should be adaptive, continuously learning from feedback, and transparent, using open data and visible algorithms. It must remain pluralistic, allowing local customization, and regenerative, enhancing ecological and social capital rather than merely optimizing flows.

This is the ethos of digital sustainability: to treat digital systems not as tools of control but as ecosystems of care, where technological advancement is harmonized with ecological and social justice.

When the Algorithm Becomes the Ecosystem

Imagine a near future where a regenerative farm earns on-chain biodiversity credits verified through drone imagery; a neighborhood app notifies residents when the grid is running on coal instead of solar, prompting shifts in energy use with small rewards; personal carbon footprints are automatically factored into e-commerce and travel decisions; and governments use AI-driven simulations to test climate policies before implementation.

In this world, climate apps do not rely on fear or guilt but blend joy, status, and collective learning. Here, algorithms function as ecosystems—dynamic, adaptive, and aligned with planetary health.

Preparing for the Next Leap

Digital permutations, when ethically designed, can make the invisible visible, shift collective behaviors, build trust across scales, seed regenerative markets, and enable emergent, entangled transformations. Yet tools alone do not transform systems. Change arises from people, culture, narratives, and the living biosphere.

Table 8.1 highlights the synergy between digital and behavioral levers. While digital tools provide the structural backbone—ensuring accuracy, transparency, and speed—behavioral levers ensure these tools resonate with human values and daily choices. The most effective climate strategies integrate both: for example, an eCarbon Card that uses blockchain for trust (digital) and gamified nudges for action (behavioral).

Next Chapter

The chapters ahead will delve into bio-based innovations—such as gene-edited low-lignin trees[84], mycelium-based packaging[85], and industrial waste valorization[86]—that reflect the logic of quantum entanglement in material form. Digital interventions, like the eCarbon Card, have shown us how information systems and behavioral nudges can reshape economic and social patterns. Yet, digital solutions alone cannot achieve full transformation unless they are complemented by tangible, regenerative approaches that address the material flows of our economy. Having explored how digital and informational systems ("bits") can guide systemic change, we now

Table 8.1 – Digital Levers vs. Behavioral Levers

Dimension	Digital Levers (Technology-Driven)	Behavioral Levers (Human-Centric)
Core Function	Data capture, automation, transparency	Habits, social norms, and identity-driven motivation
Examples	Blockchain for carbon credits, IoT sensors, AI-based forecasting	Nudges, commitment devices, peer comparisons
Feedback Speed	Real-time analytics and alerts	Gradual, through social reinforcement and learning
Scalability	Can rapidly scale through digital networks and platforms	Scales through cultural adoption and collective behavior
Trust Mechanism	Immutable records, smart contracts, algorithmic reliability	Storytelling, shared values, and community validation
Impact Focus	Structural and systemic efficiency	Behavioral shifts, lifestyle alignment with sustainability
Best Used When	Complex data and transactions need verification or automation	Individual and group choices require social or emotional cues

shift our focus to biomass and living systems—the next frontier of sustainable transformation. In the coming chapters, we turn from bits to biomass, from the realm of information systems to the tangible domain of living, regenerative materials and systems.

Chapter 9

Bio-Based Futures and Material Metamorphosis

How Biology, Circularity, and Regeneration Are Reshaping the Material World

Nature never makes waste—it only makes relationships. Sustainability begins when we remember how to make matter matter again.

"Only we humans make waste that nature can't digest"
— *Charles Moore*

The Material Foundations of the Climate Crisis

The sustainability conversation often gravitates toward energy and emissions, but beneath these lies a deeper substrate: materials—the things we build, wear, package, and consume. The global materials economy drives over 70% of global greenhouse gas emissions when one accounts for extraction, manufacturing, and disposal. More than 90 billion tons of resources are extracted annually, and this figure is expected to double by 2060. This extraction degrades ecosystems, depletes freshwater, and erodes biodiversity at unprecedented scales.

What we face is not simply a supply chain challenge; it is a metabolic crisis. Industrial metabolism remains largely linear, fossil-fed, and degenerative. A sustainable future requires a material metamorphosis—a fundamental shift from extraction to regeneration, from linear throughput to circular flows, and from inert commodities to living systems.

A New Philosophy of Matter: From Product to Process

Traditional material science has long prioritized strength, durability, cost, and scalability. While these metrics remain important, the age of ecological constraints demands that materials also optimize for relationships—with ecosystems through biocompatibility and degradability, with time through repairability and renewability, with justice through equitable sourcing and dignified labor, and with carbon through minimized embodied emissions and the potential for sequestration. Materials can no longer be treated as passive; they must be understood as active agents in systems change, embodying both function and ecological intelligence.

Examples of Material Metamorphosis

Gene-Edited Low-Lignin Trees: Decarbonizing the Pulp & Paper Industry[87]

One of the most promising, yet often overlooked, innovations in industrial decarbonization lies in forestry and fiber processing. The pulp and paper sector, which relies on energy-intensive and chemically harsh lignin-removal processes, offers fertile ground for disruption. Through CRISPR gene editing, researchers have developed low-lignin trees—particularly hybrid poplars—that significantly reduce the energy and chemical inputs needed for pulping. These trees yield more usable cellulose, grow faster, and integrate effectively with enzymatic or bio-pulping techniques.

The systemic benefits extend beyond pulp and paper. Reduced energy and chemical demands lower the industry's carbon intensity, while water and resource efficiency improve. Low-lignin feedstocks can also serve as inputs for bioplastics, cellulosic textiles, and advanced packaging materials. Moreover, lignin itself, traditionally treated as waste, can be valorized into specialty resins, carbon foams, or biofuels. This is not a mere substitution—it is a quantum-level redesign of biomass and its industrial metabolism.

Mycelium Composites: Grown, Not Made[88, 89]

Mycelium, the root network of fungi, is revolutionizing material design with its ability to grow structural composites for packaging, insulation, acoustic panels, leather-like textiles, and even architectural components. These materials are cultivated from agricultural or industrial waste—such as sawdust, deinking sludge, or paper mill residues—then molded, dried, and stabilized through heat treatment. The result is a lightweight, compostable, and moldable material that can displace fossil-based foams like polystyrene or polyurethane.

The beauty of mycelium lies in its regenerative logic. It sequesters carbon during its growth phase, turns low-value waste into high-value materials, and safely returns to the soil post-use. It exemplifies the quantum idea of entangled regeneration, where waste becomes form, and form seamlessly returns to growth.

Mycelium-Based Leather: Grown Alternatives to Animal Hides[90, 91, 92]

Companies like MycoWorks and Bolt Threads (previously) are pioneering mycelium-based leather alternatives that mimic the look, feel, and durability of animal leather without the resource intensity or animal cruelty. Using agricultural residues such as sawdust, hemp hurds, or paper mill waste as a substrate, mycelium is cultivated under controlled conditions to form dense, fibrous mats. These mats are then tanned, dyed, and finished to achieve leather-like aesthetics. Compared to traditional leather, this process reduces greenhouse gas emissions, avoids toxic chromium tanning chemicals, and enables rapid production cycles. By sourcing feedstocks locally, mycelium leather manufacturing can integrate into regional circular economies, valorizing waste streams and reducing transport emissions.

Cactus-Based Vegan Leather: Desert-Adapted Biomaterials[93]

Derived from mature prickly pear cactus leaves, cactus-based vegan leather uses minimal water and grows without pesticides on marginal land unsuitable for conventional crops. Companies like

Desserto® harvest only mature leaves, allowing the plant to regenerate naturally. The processed cactus fibers and biopolymers yield a supple, durable leather alternative used in fashion, upholstery, and accessories. With a cultivation footprint far lower than both cattle ranching and synthetic leather production, cactus leather exemplifies climate-smart agriculture and bio-based material design adapted to arid ecosystems.

Pineapple Leaf Fiber Leather (Piñatex®)[94]

Piñatex® transforms discarded pineapple leaves—a byproduct of pineapple farming—into a nonwoven mesh that is coated with a bio-based resin to create a leather-like textile. This innovation requires no additional farmland, prevents agricultural waste burning, and provides supplementary income for farming communities. The resulting material is lightweight, breathable, and strong enough for footwear, bags, and furniture. Its supply chain ties agricultural livelihoods directly to sustainable material production, closing loops between food and fashion industries.

Apple Peel-Based Vegan Leather[95]

Apple peel leather, produced from pomace and peels discarded during juice and cider production, combines plant residues with bio-based binders to create a smooth, leather-like surface. Originating in Italy, this process repurposes thousands of tons of fruit waste annually, reducing landfill methane emissions. The resulting material is waterproof, customizable in texture, and used in apparel, accessories, and interior design. By valorizing a high-moisture, otherwise perishable waste stream, apple peel leather highlights the potential of agro-industrial byproducts as premium raw materials for sustainable fashion.

Cellulosic and Regenerative Polymers: Designing for Disappearance[96]

Plastics have become the most persistent and problematic materials of the Anthropocene. Fossil-based polymers can persist for centuries, fragmenting into microplastics that contaminate ecosystems and food chains. Bio-based polymers such as polyhydroxyalkanoates (PHAs) and polylactic acid (PLA) are emerging as viable alternatives. PHAs, produced by microbial fermentation of organic feedstocks, are marine-biodegradable. PLA, derived from corn or sugarcane, is compostable under industrial conditions.

Cellulose esters—derived from wood pulp—are finding applications in filters, films, and fibers, closing loops in multiple industries. However, the challenge lies not just in sourcing sustainable feedstocks but in designing robust end-of-life pathways. Digital tagging, enzyme-assisted depolymerization, and decentralized composting networks are all necessary to ensure these polymers truly "disappear" without harmful residues.

Seaweed and Kelp to Fibers and Nonwovens[97]

Marine biomass offers one of the most exciting frontiers in sustainable material innovation. Companies like AlgiKnit and Kelsun™ (Keel Lbs) are pioneering the conversion of fast-growing kelp into biodegradable yarns, textiles, and nonwovens. Unlike terrestrial crops, kelp requires no arable land, fertilizers, or freshwater. It sequesters carbon while supporting marine biodiversity, offering a dual climate and ecological benefit.

These fibers are being developed for sustainable fashion, nonwoven medical products, and ocean-positive packaging solutions. By integrating ocean farming into supply chains, this innovation links climate mitigation with regenerative marine ecosystems—a rare but crucial intersection of economy and ecology.

Spoiled Milk to Fibers and Textiles[98]

Another remarkable innovation repurposes spoiled or expired milk, which is otherwise a source of methane emissions when discarded. Casein protein extracted from milk can be polymerized and spun into biodegradable fibers that feel silky and soft. Though casein fibers date back to the 1930s (e.g., "Lanolyn"), modern fermentation techniques and polymer science have revitalized this concept.

The process diverts waste from landfills, reduces food-related emissions, and creates circular loops for dairy byproducts. Applications range from apparel and baby clothing to hygiene nonwovens and medical dressings. Additives and nanotechnologies can enhance properties like antimicrobial resistance and durability, ensuring competitive performance alongside environmental benefits.

Corn Residue and Forestry Waste: Nonwovens and Textiles[99]

Agricultural residues, such as corn husks and stover, are increasingly being refined into fibers for biodegradable packaging, molded trays, and seed mats. Mechanical pulping and wet-laid nonwoven processes convert what would otherwise be burned—releasing methane—into functional materials that provide new revenue streams for farmers.

Forestry waste, including sawdust and bark, is now being used to produce high-tenacity Lyocell fibers through solvent spinning methods like NMMO or ionic liquid processing. These fibers outperform viscose in terms of durability and lower toxicity, while using significantly less water. A **material flow diagram** given in **Appendix XVI** illustrates this pathway: **Forestry Waste → Dissolution in Solvent → Fiber Regeneration → High-Performance Textiles**. By tapping into FSC-certified wood waste, the textile industry reduces its reliance on both cotton and synthetic fibers.

Thooshan's Biodegradable and Partially Edible Tableware[100]

Thooshan replaces single-use plastics and compostable alternatives with tableware made from agricultural by-products such as wheat bran, rice husk, rice powder, and rice bran. Their fully automated facility in Coimbatore, India, uses a dry process requiring minimal water, and produces biodegradable plates, bowls, and straws that decompose within 30 days. Unused items gracefully return to the earth as manure or serve as feed for animals like poultry and fish, championing both zero waste and circular economy principles. With government and UNDP accolades reinforcing their sustainable impact, Thooshan doesn't merely substitute plastic—it reimagines dining systems by aligning production, disposal, and ecosystem health in regenerative loops.

Plastic-Eating Wax Moth Larvae[101, 102, 103, 104, 105]

Federica Bertocchini's accidental discovery of wax moth larvae's plastic digestion capability has helped the world to recognize that biologically based recycling can be an answer to environmental pollution, and the Earth's ecosystem has a tremendously useful "ecosystem service" that can tackle the global plastic trash catastrophe that is harming global ecosystems. In 2017, Bombelli et al. (2017)[106] published a brief report showing the wax moth larvae (Galleria mellonella)'s ability to degrade Polyethylene, the most common packaging plastic, rapidly.

Coffee-Constituted Concrete[107, 108]

Disposal of organic waste can generate significant greenhouse gas emissions, including methane emissions from anaerobic fermentation that produces landfill gas. In 2023, a team of researchers and engineers at RMIT University discovered how to recycle coffee grounds to produce construction materials instead of letting them decompose and emit greenhouse gases. This method not only helps to create values for an estimated 508 million kilograms of used coffee grounds a day globally but also offers hope for recycling other types of organic materials generated as byproducts from agricultural activities and food production processes. The discovery also helps reduce the pressure on sand and gravel exploitation and strengthen the concrete by roughly 30%[109].

Carbon-Sequestration Algae[110, 111, 112, 113, 114, 115, 116]

Numerous solutions have been proposed to capture and sequester carbon to confront the climate and environmental degradation crises. One of the most crucial and effective methods today is the biosequestration of carbon through algae. Both macroalgae and microalgae show significant potential in reducing greenhouse gas emissions, mitigating ocean acidification, and contributing to national energy security[117]. Additionally, carbon capture and sequestration through algae are viewed as environmentally friendly and economically feasible and contribute to sustainability in the long term[118, 119, 120, 121].

Table 9.1 – Material Futures Matrix

Innovation	Feedstock / Source	Carbon Impact	Circularity Potential	Scalability	Key Applications
Gene-Edited Low-Lignin Trees	Hybrid poplar, engineered forestry biomass	Significant reduction in pulping emissions; lower energy and chemical use	High – residual lignin valorized into resins or biofuels	Medium–High (requires forestry adoption and regulation)	Paper, fiberboard, bioplastics feedstock
Mycelium Composites	Agricultural/industrial waste (sawdust, deinking sludge)	Carbon sequestration during growth; replaces fossil foams	Fully compostable; waste-to-growth cycle	Medium (growing manufacturing capacity)	Packaging, insulation, acoustic panels, textiles
Mycelium-Based Leather	Sawdust, hemp hurds, paper mill residues	Avoids livestock emissions and toxic tanning chemicals	High – can valorize local waste streams	Medium (scaling biofabrication facilities)	Fashion, upholstery, accessories
Cactus-Based Vegan Leather	Prickly pear cactus leaves	Minimal water, no pesticides; avoids livestock emissions	Renewable cultivation; biodegradable with proper systems	Medium (regional adaptation)	Fashion, upholstery, accessories
Piñatex® – Pineapple Leaf Leather	Pineapple leaves (agricultural waste)	Avoids burning waste; offsets synthetic and animal leather	High – utilizes byproduct of existing farming	Medium (seasonal, regional limits)	Footwear, bags, furniture
Apple Peel Vegan Leather	Apple pomace and peels	Diverts food waste; avoids methane emissions	Renewable feedstock; biodegradable with proper binders	Low–Medium (specialized processing)	Apparel, accessories, interiors
Cellulosic Polymers (PHA/PLA)	Corn, sugarcane, microbial fermentation	Reduced lifecycle emissions vs. petro-plastics	Compostable or enzyme-degradable with proper systems	High (existing industrial capacity)	Films, fibers, packaging, filters

Seaweed Fibers (e.g., Kelsun™)	Kelp and brown algae	Net carbon sink during growth; no land use	Compostable; ocean farming closed loop	Medium (scaling sea farms)	Apparel, nonwovens, medical textiles, packaging
Spoiled Milk Fibers	Expired dairy milk (casein protein)	Diverts food waste; reduces methane emissions	Biodegradable, circular dairy valorization	Low–Medium (needs local dairy waste streams)	Apparel, hygiene textiles, medical dressings
Corn Husk Nonwovens	Corn residue and stover	Avoids methane from burning; replaces plastics	Fully biodegradable; boosts rural income	Medium (seasonal feedstock)	Seed mats, molded trays, packaging
Forestry Waste Lyocell	Sawdust, bark (FSC-certified)	Lower emissions than cotton/ viscose	Closed-loop solvent spinning; high recycling potential	High (wood industry integration)	Apparel, workwear, wipes, durable textiles
Citrus Peel Leather	Fruit-processing waste	Avoids landfill emissions; displaces animal leather	Renewable, biodegradable resin matrix	Low–Medium (specialized processing)	Vegan leather products, footwear
Thooshan Biodegradable Tableware	Wheat bran, rice husk, rice powder, rice bran	Replaces single-use plastics; composts in 30 days	Zero-waste loop; can become feed or manure	High (existing automation)	Plates, bowls, straws
Plastic-Eating Wax Moth Larvae	Galleria mellonella larvae	Biodegrades polyethylene waste	Potential biological recycling pathway	Low (requires containment, scaling research)	Plastic waste treatment
Coffee-Constituted Concrete	Spent coffee grounds	Diverts organic waste; strengthens concrete ~30%	Integrates organic byproduct into construction	Medium (requires collection systems)	Construction materials
Carbon-Sequestration Algae	Macroalgae and microalgae biomass	Captures CO_2; mitigates ocean acidification	Renewable; bio-based carbon storage	Medium–High (mariculture scalability)	Biofuels, bioplastics, carbon offsets

Waste as Substrate: Industrial Ecology and Circular Manufacturing

The best bio-based innovations start not with virgin biomass but with waste. Pulp mill sludge can serve as a mycelium substrate, citrus peels can be transformed into vegan leather, and brewery grains can be upcycled into biochar or protein feed. Lignin streams—once a low-value byproduct—are now being converted into adhesives, carbon foams, or even supercapacitor materials. This is not conventional recycling. It is an industrial ecology where waste is a nutrient, feeding multi-product value chains that mimic fungal metabolism—connective, opportunistic, and regenerative.

Table 9.1 above summarizes and compares the key bio-based innovations discussed. It highlights their carbon impact, scalability, and potential for circularity while retaining narrative clarity.

<u>Key Insights</u>

- **Carbon Impact** – Mycelium composites and seaweed fibers offer not just low emissions but actual carbon sequestration during growth.
- **Circularity** – Forestry waste Lyocell and citrus peel leather exemplify high circularity when paired with local collection and closed-loop processes.
- **Scalability** – Gene-edited low-lignin trees and cellulosic polymers have the highest potential for large-scale impact, though adoption barriers differ—forestry requiring regulatory approval, and polymers requiring composting infrastructure.

Quantum Metabolism: Designing Entangled Material Systems

Quantum principles provide a fresh lens for rethinking material systems. Superposition suggests that one feedstock can serve multiple roles—food, fiber, or fuel—depending on context. Entanglement reminds us that material flows are interconnected, linking agriculture to packaging, fashion to energy. Decoherence occurs when systems lose integration and materials fall out of circulation, while emergence arises when regenerative feedback loops produce unexpected innovations. Just as quantum particles behave relationally, sustainable materials must be co-designed with the ecological and social systems they inhabit.

Challenges and Cautions

Bio-based materials are not a panacea. Without careful governance, they can lead to land-use conflicts when agricultural land is diverted to biomass crops, industrial greenwashing through misleading biodegradability claims, and energy-intensive biorefineries that negate climate gains. Monocultures of fast-growing feedstocks may harm biodiversity and soil health. True sustainability must evaluate not just the molecular structure but the entire system surrounding the material.

A Framework for Material Transformation

To guide innovation, policy, and investment, the **Material Metamorphosis Framework** proposes six dimensions: sourcing from regenerative or waste biomass; using low-energy, low-toxicity production methods; ensuring functionality that meets or exceeds conventional alternatives; designing for circular end-of-life pathways (composting, recycling, or upcycling); integrating materials into entangled, circular flows across sectors; and ensuring equitable value distribution along the supply chain.

Designing for Dignity and Decarbonization

The material revolution must also be a social revolution. Rural and coastal communities can become stewards and producers of regenerative materials, while decentralized biomanufacturing enhances regional resilience. Traditional knowledge—such as bark cloth, coir, and hempcrete—can be revalorized and blended with cutting-edge biotechnology. Jobs in mycelium fabrication, textile upcycling, and circular logistics offer not just economic opportunities but a chance to decentralize prosperity and build dignity into industrial systems. Avoiding the extractive, inequitable models of fossil capitalism is as important as reducing emissions.

From Dead Matter to Living Systems

Bio-based innovation marks a paradigm shift. Materials cease to be inert, linear commodities and instead become agents of regeneration. This shift moves us from throughput logic to metabolic logic, from isolated products to interconnected systems. As quantum mechanics reveals the relational nature of matter, sustainable materials must be seen as part of living networks, designed not to dominate nature but to collaborate with it.

Transition to Energy Systems as Living Systems

The next chapter extends this philosophy to energy. Just as materials can embody circularity and life, so too can energy systems be designed as decentralized, adaptive, and self-organizing networks—mimicking the resilience of ecosystems. We now move from bio-based to grid-based innovation, from matter to flow, while holding onto the central idea: in a quantum world, everything is entangled, and so too must our solutions be.

Chapter 10

Energy Systems as Living Systems

Resilience, Adaptation, and Decentralization in a World of Flow

> If matter is memory, then energy is intention. To live sustainably is to flow wisely.

Reframing Energy: From Machine to Metabolism

For over a century, humanity has treated energy systems as machines—predictable, centralized, optimized for throughput, and controlled from the top down. This machine model powered industrial growth but came at a steep cost: gigaton-scale carbon emissions, systemic fragility, exclusion of billions from affordable access, and the dumping of externalities into air, water, and politics. To decarbonize the energy system is not merely to change fuels; it is to change metaphors. We must begin to see energy not as an input but as a flow—alive, adaptive, and entangled—much like an ecosystem.

Energy as Ecosystem: Key Principles

If we were to design energy systems as living systems, they would embody the characteristics of ecosystems: decentralization, adaptation, circularity, redundancy, and equity. Just as ecosystems distribute biomass across niches, energy should be locally sourced and resilient. Rooftop solar, microgrids, community wind, and bioenergy hubs create a diverse energy ecology rather than a hierarchical pyramid. Adaptation would become central, with systems learning and responding to weather, demand, and supply in real time, guided by smart meters, AI-driven controls, and blockchain contracts that act like the nervous and endocrine systems of an energy organism.

Circularity would be fundamental: waste heat captured for district heating, EV batteries repurposed for home storage, and every energy output cycled rather than discarded. Redundancy and modularity would ensure there is no single point of failure; backup systems and diversity of sources—solar, wind, geothermal, hydro, hydrogen—would function much like the biodiversity that makes coral reefs resilient. Think coral reefs—not pipelines[*] (Fig. 10.1).

Equity must also become a design feature rather than a byproduct. Energy justice—access, affordability, and agency—should form the root system of the new energy forest. In nature, energy is captured and reused locally. Distributed energy systems, particularly microgrids with bidirectional flow, mimic this ecological intelligence. They reduce transmission losses, build local autonomy, and adapt to the rhythms of sun, wind, and biomass specific to each place.

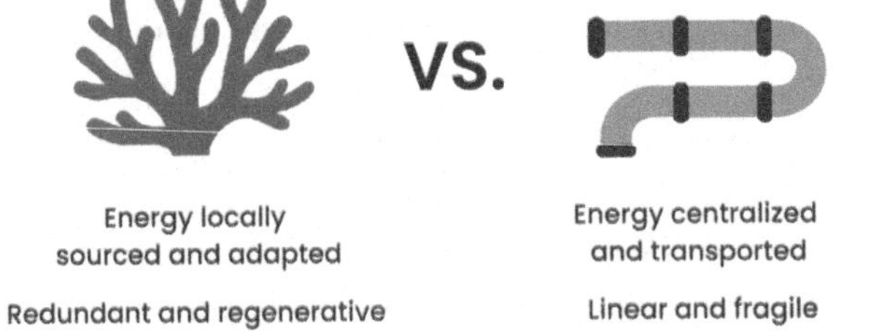

Figure 10.1

* "Think coral reefs—not pipelines" is a call to shift from centralized, single-purpose, extractive systems to decentralized, multi-functional, and regenerative ones. Coral reefs represent adaptive design embedded in ecological intelligence. In a decarbonized future, infrastructure must behave more like living systems—capable of healing, adapting, and co-evolving with their environments.

Failures of the Centralized Paradigm

Today's energy systems remain largely centralized around thermal power plants, transmitted through brittle grids and optimized for profit rather than equity or regeneration. They are vulnerable to blackouts from climate events—such as the 2021 freeze in Texas—inefficient due to transmission losses, blind to demand-side agency where the consumer remains a passive endpoint, and slow to adapt to distributed renewable generation. This centralized paradigm is a Newtonian relic of a linear age, ill-suited to the nonlinear shocks and uncertainties of the Anthropocene.

The Emergence of Living Grids

Across the globe, a new model is emerging: the living grid. These systems move from centralized fossil-heavy generation to distributed, renewable, and modular energy flows. Transmission becomes two-way and localized, while AI-assisted optimization replaces rigid, hierarchical control. Users evolve from passive consumers to active prosumers—producers and consumers who store, share, and manage energy at the edge of the grid. Living grids measure success not by throughput or profit but by resilience, equity, and circularity.

Living grids operate much like forest ecosystems. Solar rooftops function like leaves capturing sunlight, electric vehicles become energy-storing fruits, batteries serve as roots storing potential energy, and AI acts as a mycorrhizal network sensing and routing flows dynamically across the system. Table 10.1 provides a comparison summary of the Centralized vs. Living Energy Systems.

The Role of AI and Quantum Computation

Complex, decentralized energy systems cannot be effectively managed alone through traditional spreadsheets, static models, or purely linear forecasts only. As energy ecosystems become increasingly dynamic—with millions of interconnected nodes spanning

Table 10.1 – Comparative Table: Centralized vs. Living Energy Systems

Dimension	Centralized Grid (Mechanical)	Living Grid (Ecosys-temic)
Generation	Centralized, fossil-heavy	Distributed, renewable, modular
Transmission	One-way, long-distance	Two-way, localized + regional
Control	Manual or hierarchical	AI-assisted, self-optimizing
User Role	Passive consumer	Active prosumer and network participant
Resilience	Fragile, prone to large disruptions	Redundant, nested, adaptive resilience
Metrics of Success	Profit, throughput	Resilience, equity, circularity

rooftop solar panels, battery storage systems, electric vehicles, and smart appliances—the need for advanced computational intelligence becomes paramount. Artificial Intelligence (AI), machine learning (ML), and, potentially, quantum computing offer unprecedented capabilities to handle this complexity. These tools can forecast solar irradiance and wind velocity with high precision, model dynamic pricing and demand-response behaviors, detect faults in real time, reroute energy flows seamlessly, and optimize usage across distributed networks. Importantly, these technologies are not designed to impose rigid control but to facilitate emergent patterns of coordination—much like a conductor guiding the tempo and tone of a symphony without dictating every note. These tools don't control—they **facilitate emergence**.

A useful analogy can be drawn between energy systems and mobility platforms. Traditional, centralized power grids resemble legacy taxi dispatch systems—hierarchical, rigid, and often slow to adapt to fluctuating demand. In contrast, ride-sharing platforms such as Uber operate on a decentralized, demand-driven model that dynamically matches riders with drivers, continuously reroutes vehicles to optimize efficiency, and leverages real-time data to minimize delays. Distributed energy systems (DES) are evolving in a similar way. By digitally orchestrating peer-to-peer energy flows and dynamically matching supply with demand, they achieve greater flexibility and resilience. Modular energy units—such as rooftop solar arrays, home batteries, or community microgrids—can be seamlessly integrated and optimized, reducing the risk of cascading failures that plague centralized systems.

Looking further ahead, the future of energy management may resemble the complexity of emerging air mobility platforms—eVTOL (electric vertical takeoff and landing) air taxis or swarm-coordinated drones—that optimize routes, avoid disruptions, and adapt to changing conditions in real time. Just as these aerial systems rely on distributed intelligence and rapid coordination, decentralized energy ecosystems will leverage AI and quantum-inspired computation to route power, anticipate surges, and enhance stability. These technologies, by design, will transform the energy grid into a living, adaptive network—capable of self-healing, responding to unpredictable events, and continuously improving efficiency while empowering prosumers and communities.

Entangled Energy Systems

Energy does not exist in isolation; it is deeply intertwined with water, food, mobility, carbon, and equity*. Every energy choice we make reverberates across these domains, creating ripple effects that can either reinforce or undermine sustainability. For instance, thermal power plants require vast amounts of freshwater for cooling, often competing with agriculture and urban water needs, especially in water-stressed regions. Conversely, pumping, treating, and distributing water is highly energy-intensive, forming a tightly coupled energy-water nexus. Agriculture further illustrates this entanglement:

*This scenario-based framework (lens) – a subset of the SCI framework, demonstrates how interconnected domains like mobility and carbon align with SCI categories—water, food, and carbon fall under planetary boundaries, while mobility integrates ecological integrity with innovation, and equity is embedded in SCI's justice lens. See Appendix XVII Table XVII.2 for cross-alignment with SCI.

fertilizers, irrigation, and mechanized farming are energy-intensive processes, while energy crops such as corn for biofuels compete with food production, affecting both prices and land use.

Mobility adds another layer of complexity. The rapid adoption of electric vehicles (EVs) reshapes grid demand curves, requiring sophisticated load balancing and energy storage solutions. At the same time, the decarbonization of grids is essential for EVs to deliver meaningful climate benefits, making electricity generation the linchpin of all net-zero strategies. Moreover, energy poverty—still affecting hundreds of millions globally—exacerbates disparities in health, education, and livelihoods, limiting access to opportunities that depend on reliable and affordable power.

Solutions must therefore account for these systemic entanglements rather than address energy as a standalone challenge. For example, pairing EV rollouts with time-of-use tariffs, home solar systems, and vehicle-to-grid (V2G) technology can reduce grid stress during peak demand while maximizing environmental and economic benefits. Similarly, integrating **waste-to-energy** systems within urban sanitation infrastructures not only addresses hygiene and waste disposal but also generates energy and fosters dignity for underserved communities. Renewable-powered microgrids, when combined with water recycling systems, can simultaneously tackle energy access, water scarcity, and resilience in disaster-prone areas.

Energy transitions, when viewed through this lens of entanglement, become far more than technical fixes. They evolve into platforms for holistic transformation—bridging equity gaps, improving resource efficiency, and linking diverse sectors into a more coherent and regenerative whole.

Digital-Physical Hybrids: The Rise of the "Electro-Ecological Stack"

A new architecture for energy and sustainability is emerging—one where physical infrastructure seamlessly converges with digital intelligence to form a highly adaptive, decentralized ecosystem. At its foundation, the physical layer consists of rooftop solar arrays, microgrids, wind turbines, biogas digesters, and green hydrogen systems, all operating as modular, distributed units that can scale up or down as needed. These decentralized systems reduce dependence on centralized grids, improve resilience to disruptions, and enable local energy sovereignty.

Overlaid on this physical layer is a digital layer, which brings intelligence and coordination to the network. IoT sensors continuously monitor energy generation, storage, and consumption. Blockchain-enabled smart contracts facilitate transparent, trustless energy transactions—allowing households, small businesses, and even electric vehicles to trade surplus energy in peer-to-peer markets. AI-driven optimization platforms dynamically balance loads, forecast demand, and manage energy flows in real time, ensuring that renewable energy sources are integrated efficiently into the grid.

The social layer adds a cooperative dimension. Energy cooperatives, local utilities, and user-owned data commons empower communities to govern their energy systems democratically. For instance, neighborhood solar collectives or rural energy cooperatives in India and Africa are proving that decentralized energy systems can be both inclusive and profitable, fostering local entrepreneurship and skill development.

On top of these layers lies the financial layer, which is redefining how value is created and exchanged. Tokenized incentives and carbon-linked micropayments reward behaviors such as energy saving, load shifting, and carbon offsetting. Decentralized energy markets leverage these tokens to price energy in real time, aligning financial signals with sustainability goals.

This multi-layered integration—physical, digital, social, and financial—forms what can be described as an "Electro-Ecological Stack." It is not governed by a single authority but is choreographed by distributed intelligence, with each node in the system contributing to collective resilience and optimization. Much like a quantum system, the electro-ecological stack thrives on **entanglement and feedback**, where every local decision can ripple through the network to optimize global outcomes.

The Village and the Grid: India's Energy Quantum Leap

India offers one of the most compelling examples of how energy transitions can become **entangled opportunities** that drive social, economic, and environmental transformation simultaneously. With over 300 sunny days a year, a vast population living in rural areas, and high mobile phone penetration enabling digital coordination, India stands at the cusp of an energy revolution uniquely tailored to its context. The country's rural electrification challenge, long seen as an infrastructure gap, is increasingly being reframed as a chance to leapfrog directly into distributed, clean, and intelligent energy systems.

The combination of India Stack, Aadhaar (digital identity infrastructure), and the Unified Payments Interface (UPI) provides a ready-made platform for integrating digital identity, micro-payments, and energy services. This convergence opens the door for reinventing energy subsidies—traditionally blunt instruments prone to inefficiency—into precision-targeted, behavior-linked incentives. Farmers, for instance, could receive energy credits tied to low-carbon agricultural practices, while urban and rural households could track and trade their energy savings in real time.

Imagine a solar-powered microgrid in a rural village, fully integrated with AI-powered load forecasting and dynamic pricing that balances energy supply and demand. Residents track their energy usage via mobile apps, while eCarbon Cards reward low-carbon behavior, from efficient appliance use to surplus solar exports. Excess energy is seamlessly traded with neighboring villages through blockchain-enabled smart contracts, ensuring that energy flows not only meet local needs but also build economic value within the community. Such a system creates energy democracy, where villagers are not just passive consumers but active producers, traders, and custodians of energy resources.

This decentralized approach is more than a technical fix; it is intertwined with development, dignity, and decarbonization. It supports local job creation—through solar installation, maintenance, and digital services—while reducing dependence on centralized fossil-based grids. Moreover, it aligns with India's national decarbonization targets and its ambition to become a leader in renewable energy and green hydrogen. In this sense, India's villages could become global prototypes of distributed energy futures, showcasing how quantum-inspired, entangled energy ecosystems can achieve both sustainability and equity.

Challenges: Not All That Is Green Is Just

Despite the transformative promise of decentralized and renewable energy systems, they are not inherently just or equitable. Large-scale solar farms, for example, have displaced Indigenous communities and traditional land rights in several regions, amounting to a new form of "green colonialism." Similarly, the extraction of lithium, cobalt, and rare earth minerals—critical for batteries and renewable technologies—brings with it environmental degradation, labor exploitation, and geopolitical tensions. These "clean" technologies can externalize harm to vulnerable communities, often far removed from the benefits they provide.

The digital layer of energy transitions also raises risks. Data monopolies may emerge if technology giants dominate energy platforms, controlling both data and market access. Off-grid and marginalized regions may be left behind in the push for smart, connected grids, deepening the energy divide rather than bridging it. A just energy transition must therefore go beyond decarbonization to address power dynamics—who decides, who benefits, and who participates—ensuring that no community is sacrificed for the sake of progress.

Design Principles for Energy as a Living System

If energy systems are to behave like living organisms, they must be modular, regenerative, and adaptive. Modular systems allow local components—such as village microgrids—to grow, evolve, or fail gracefully without destabilizing the whole. Real-time feedback loops, powered by IoT and AI, must enable continuous learning and rapid responses to shifting conditions, replacing static, top-down reporting.

Governance, too, should be distributed. Communities should own, govern, and profit from their energy resources, turning energy systems into engines of local empowerment rather than centralized control. Nested scales are essential: village microgrids must seamlessly connect with urban grids and national networks, creating a multi-layered ecosystem of energy flows that mirrors the logic of natural systems.

Energy should follow temporal coherence, aligning with daily and seasonal cycles such as sun paths and wind patterns, while adopting a regenerative logic that restores ecosystems, strengthens economies, and uplifts communities, rather than depleting them.

The Soul of Energy

At its heart, this chapter is not simply about grids, gigawatts, or technical efficiency. It is about reimagining our relationship to power—both as electricity and as agency. Energy defines how we live, act, and contribute. Do we approach it with a mindset of consumption or contribution, extraction or regeneration, dominance or harmony?

To treat energy as a living system is to recognize that sustainability is not a static end-state but a practice of alignment and reciprocity. Just as a healthy ecosystem thrives through interdependence, a just energy transition must weave together technology, community, culture, and ecology into a coherent whole. Energy, in this sense, becomes not just a commodity but a shared rhythm and a collective heartbeat.

What Comes Next

In the chapters that follow, we shift our focus to finance and governance—the invisible architectures that determine how energy, innovation, and resources flow. We will explore quantum-aware carbon indices and market feedbacks that redefine risk, value, and opportunity. We will examine polycentric governance structures that entangle local and global priorities while using digital traceability to foster transparency and trust.

Finally, we will look at leadership and citizenship, where individuals and communities become quantum actors driving systemic transformation from the ground up. Our energy walk now leads to capital and regulation—the currency of coordination—but our guiding principle remains constant: **sustainability is not imposed; it emerges, entangles, and evolves.**

As we consider the architectures of finance, governance, and leadership that will determine how energy transitions unfold, it becomes clear that our choices today will reverberate across decades. Yet abstract discussions of capital flows and regulatory frameworks can feel detached from lived experience. To bridge this gap, we now turn to **future scenarios**—imaginative but plausible snapshots of how energy systems might look in 2035 if different pathways of technology, culture, and governance entangle productively. These vignettes are not predictions but **provocations**: stories designed to help us think beyond incremental change and to experience what systemic transformation might feel like on the ground—in villages, cities, deserts, and coastal commons.

Future Scenario Snapshot | Global Energy Futures Gallery (2035 Snapshots)

Rural (Living) Energy Commons

It is the year 2035. In a mid-sized Indian village, energy is no longer something extracted and purchased—it is co-created, shared, and stewarded like a common resource. The rooftops glisten with solar panels, collectively owned by the community cooperative. Each home, equipped with modular battery packs repurposed from second-life EV batteries, acts as both a consumer and a contributor, storing energy during the day and sharing excess power at night through a digital peer-to-peer marketplace.

The local microgrid hums with distributed intelligence. A quantum-enhanced AI, trained on years of weather, demand, and agricultural cycles, continuously balances flows between the village and the nearby town, ensuring that power is allocated not just for efficiency but for equity. Critical infrastructure—schools, health clinics, and cold storage for farmers—receives guaranteed baseline energy, while flexible loads like irrigation pumps and community halls operate on dynamic tariffs guided by surplus generation.

Residents engage with the system through a simple mobile interface. The app, integrated with their eCarbon Cards, tracks personal and household energy use in real time, rewarding low-carbon behaviors with micro-incentives redeemable for local services or cooperative dividends. Instead of seeing energy as a monthly bill, they now see it as a form of participation—an ecosystem to which they belong.

Waste heat from a nearby agro-processing unit is funneled into a district heating network that dries grains and warms community centers during winter evenings. Biogas digesters fed by kitchen waste provide fuel for communal cooking events. Energy decisions are not made in distant boardrooms but in village assemblies, where residents, guided by digital dashboards and storytelling, shape the rhythms of the grid.

This energy commons is not just about technology; it is about culture. The shift to decentralized, adaptive energy has revitalized community bonds. Young people, once migrating to cities for stable jobs, are now trained as local energy stewards—managing digital platforms, maintaining equipment, and experimenting with small-scale innovations like vertical wind turbines or algae-based biofuel pods. Elders share traditional weather knowledge, blending it with satellite forecasts to anticipate seasonal shifts.

What was once a fragmented energy supply chain—dominated by distant utilities and volatile fossil fuel prices—has evolved into a living, learning organism. The system flexes and breathes like a forest, adapting to storms, embracing redundancy, and restoring balance when stressed. Energy poverty has become energy prosperity, not through handouts but through collective intelligence and design.

By 2035, this model has spread across regions, connecting hundreds of microgrids into resilient "energy webs" that exchange power and information seamlessly, much like mycorrhizal networks beneath the soil. These networks are not just technical infrastructures but cultural commons—spaces where energy is intertwined with dignity, livelihood, and ecological care.

Urban Energy-as-a-Service: New York's Quantum Grid District

In a revitalized borough of New York, the concept of energy ownership has dissolved into a seamless subscription-based service. Residents no longer receive electricity bills but participate in a collective "energy-as-a-service" model. A network of rooftop solar, community wind turbines, and underground thermal storage is orchestrated by a quantum AI that forecasts urban demand patterns with extraordinary precision. Each building operates as a prosumer node, automatically trading energy surpluses through blockchain-enabled smart contracts.

EV fleets double as mobile energy storage units, parking near schools and hospitals during peak times to supply backup power. Residents interact with a single digital platform that manages everything—energy use, carbon credits, and maintenance schedules—while rewarding low-carbon lifestyles with micro-incentives. Streetlights dim or brighten dynamically, heating and cooling systems adapt to real-time weather shifts, and building energy flows mimic the autonomic intelligence of a living organism. The city grid no longer merely delivers power—it behaves as an ecosystem, anticipating, adapting, and self-healing when stressed.

Desert Hydrogen Oasis: Africa's Green Energy Hub

In North Africa's Sahara fringes, vast solar arrays and concentrated solar power (CSP) plants shimmer in the desert sun. Instead of exporting fossil fuels, the region has become a hub for producing green hydrogen, powered entirely by renewable energy. Excess solar energy during the day is diverted into large-scale electrolysis plants, producing hydrogen for local industry and export via ammonia carriers.

This hub is more than a cluster of energy infrastructure; it is an integrated water-energy-food nexus. Desalination plants powered by the same solar arrays provide freshwater for nearby desert farming cooperatives. Vertical greenhouses grow food under protective solar canopies, while micro-hydrogen storage systems power off-grid refrigeration for rural villages. The surplus heat from CSP units supports algae-based biofuel production, creating a cascade of energy and food benefits. Communities that were once marginalized due to water scarcity and energy poverty now thrive, exporting not just hydrogen but also expertise in circular, desert-adapted design.

Floating Solar-Fishery Hybrid: Southeast Asia's Blue Commons

In coastal Southeast Asia, where rising seas threaten livelihoods, floating solar farms have merged with traditional aquaculture to create hybrid blue commons. Solar panels float on modular rafts that also provide shade and temperature regulation for fish farms below. This integration reduces water evaporation and algae overgrowth while producing clean energy for local communities.

Blockchain-enabled microgrids allow coastal villages to trade surplus energy with mainland cities, while automated sensors monitor fish health, water quality, and solar efficiency. Excess energy is used to power cold storage facilities and desalination units, closing the loop between energy, food, and water security. Fisherfolk, once vulnerable to the volatility of climate and markets, are now active stakeholders in this multi-layered, regenerative system. The fusion of aquaculture and photovoltaics exemplifies a design ethic where infrastructure serves both human needs and ecological renewal.

Conclusion of the Gallery

These scenarios—rural energy commons, urban quantum grids, desert hydrogen hubs, and floating solar-fishery systems—are not distant science fiction. They represent the entangled potential of technology, culture, and ecology when designed as living, adaptive systems. By 2035, energy is no longer a passive commodity but a generative force, woven into the rhythms of place, people, and planet.

Chapter 11

Finance as a Field Function

Rethinking Capital Flows Through Entangled Value and Quantum Feedback

Money is not a thing—it's a signal. And in a quantum world, the strength of your signal defines the direction of your system.

Capital as Code: From Flow to Field

Finance is more than a reservoir to be drawn upon; it is an invisible field, shaping trajectories long before any transaction is made. Like a magnetic or gravitational field, capital exerts influence through its distribution, weighting, and timing. The pattern—not just the quantity—of investment determines how societies, ecosystems, and markets evolve. A single infusion of capital can reorganize supply chains, change social norms, and activate reinforcing feedback loops.

In sustainability, the architecture of finance dictates the architecture of possibility. Public, private, philanthropic, and citizen-led capital can be "re-coded" to behave less like a mechanical pump and more like a responsive, distributed, quantum field—adapting in real time, entangled with multiple outcomes, and capable of triggering transformation rather than just funding it.

The Problem with Linear Finance: Platforms of Accountability

Our prevailing financial systems are built on Newtonian assumptions—predictability, control, and linear cause-effect. They prize efficiency as the highest virtue, assume that minimizing risk ensures performance, and measure success in quarterly returns, EBITDA, or

IRR. This mindset rewards extractive, high-speed capital flows and undervalues regenerative, long-horizon projects such as deep decarbonization, ecosystem restoration, or community resilience.

Externalities—social, ecological, or intergenerational—are often excluded from the balance sheet entirely. This results in capital flows that run counter to planetary stability.

However, embedded climate feedback systems are beginning to shift this pattern. Carbon credit-linked consumer finance tools—like Sweden's **DO Black Card**[122, 123] and India's West Bengal **Carbon Credit Card**[124, 125] pilot—are introducing accountability at the point of transaction. By integrating real-time carbon data into spending, these instruments make invisible impacts visible, turning everyday purchases into moments of climate decision-making.

A Quantum Theory of Value: New Institutions and Instruments

In a quantum view, value is relational, emergent, and probabilistic. It is created through entanglement—between investors, communities, ecosystems, and time horizons—and often becomes visible only when systems reorganize themselves. A small, local investment in community solar, for example, can spark national policy shifts or cross-sector collaborations.

To reflect this complexity, financial instruments must embed:

- **Nonlinear causality** – effects disproportionate to inputs
- **Delayed returns** – ecological payback periods
- **Multi-scalar co-benefits** – local, regional, global

Table 11.1 contrasts traditional finance with quantum finance, showing how metrics, risk approaches, and narratives shift when capital is treated as a field rather than a static resource.

Micro-Infrastructures for Climate Governance

In this field-oriented model, tools like carbon-limiting credit cards, personal carbon allowances, and community energy credits become micro-nodes in a global network of accountability. Sweden's

Table 11.1 – Quantum Finance Instruments vs. Traditional

Aspect	Traditional Finance	Quantum Finance
View of Capital	Static resource; focused on accumulation	Dynamic field; capital as a signal shaping systemic outcomes
Time Horizon	Short-term returns (quarters/years)	Long-term regeneration (decades/generations)
Metrics of Success	ROI, EBITDA, IRR	Integrated Return on Interdependence, Network Potential Value
Risk Approach	Minimize and transfer risk	Redistribute, transform, and hedge risk through entanglement
Instrument Design	Single-outcome tools (bonds, stocks)	Entangled instruments (green bonds, outcome-linked loans)
Feedback Loops	Static performance reviews	Real-time adaptive feedback (AI, carbon indices)
Inclusion & Agency	Centralized control by financial institutions	Community-led capital (peer-to-peer lending, cooperative funds)
Carbon Accountability	Treated as an externality	Embedded into transactions (e.g., eCarbon Card, DO Black Card)
Narrative Power	Profit-driven storytelling	Values-driven storytelling, confidence as a regenerative force

DO Black imposes real-time carbon caps; Switzerland's **Climate Credit Card** links purchases to verified offsets; India's West Bengal pilot integrates credits into public services and tax systems.

These innovations move finance from a detached observer to an active participant in climate governance—transforming capital into a living feedback mechanism.

Rewiring Financial Infrastructure

A field function requires dynamic, distributed, and adaptive capital systems. Five strategies illustrate this transformation:

- **Recursive Finance** – Capital loops back into its source system, funding further improvements. Recursive Finance creates adaptive loops, where financial outputs feed back into future inputs. For example, revenue generated from carbon taxes can fund renewable energy transitions, dividends from sustainable enterprises can seed local circular economies, and data from impact investments can improve predictive models. The **eCarbon Card**, in this sense, represents a pioneering form of recursive finance, linking household-level carbon savings to tokenized rewards or rebates that reinforce further low-carbon behaviors

- **Entangled Instruments** – Multi-objective tools (e.g., green bonds with biodiversity clauses) that deliver layered outcomes. Entangled Instruments integrate multiple goals within a single tool. Green bonds, for example, deliver stable financial returns while reducing emissions and improving infrastructure resilience. Outcome-based loans offer lower interest rates when biodiversity or equity targets are met. Blended finance structures combine public de-risking with private capital to accelerate the adoption of frontier technologies.
- **Long-Time Capital** – Patience and regeneration replace rapid turnover. Long-Time Capital emphasizes patience and regenerative horizons. Deep decarbonization, afforestation, or coastal resilience infrastructure unfold over decades, not quarters. Sovereign climate funds—particularly in the Global South—are stepping into this role by investing in ecosystem services such as mangroves or wetlands, which generate ecological, social, and economic value for generations.
- **Feedback-Responsive Capital** – AI-linked metrics and smart contracts allow capital terms to adjust in real time. Feedback-Responsive Capital adapts dynamically rather than operating on static metrics. AI and machine learning can track ESG signals in real time, while smart contracts adjust terms automatically based on impact. Carbon-indexed instruments could tie payouts directly to emission-reduction benchmarks. The proposed **U.S. Carbon Allowance Index (USCA)**[126] is an example of this principle. By integrating data from California's Cap-and-Trade Program[127], RGGI[128], and Washington's Climate Commitment Act[129], USCA can serve as a benchmark for green investments,

a hedge against fossil risks, and a signal to synchronize capital flows with regulatory shifts.

- **Community-Led Capital** – Decentralized financing empowers local ownership and resilience. Community-Led Capital Architectures decentralize finance, allowing capital to circulate among citizens rather than flowing exclusively through banks or large institutions. Examples include peer-to-peer lending for rooftop solar projects, participatory budgeting for green infrastructure, or cooperative financing models for agroforestry ventures. In South India, farmer collectives have built bio-input centers through local revolving credit schemes, which reduce chemical dependency, regenerate soil health, and strengthen community trust.

Quantum Metrics for Quantum Markets

When finance is a field, the measures of success must also be multidimensional. Quantum-aware metrics focus on interdependence and resilience rather than isolated returns. ROI becomes an *integrated return on interdependence*, payback periods evolve into *ecological regeneration periods*, and net present value transforms into *network potential value*. Similarly, creditworthiness expands into community resilience scores, while carbon per dollar must be measured in terms of both embodied emissions and social co-benefits. Measurement must be real-time, participatory, and multi-scalar. These metrics account for resilience, adaptability, and social equity alongside financial returns.

Central Banks, Sovereign Funds, and Field Stewardship

Central banks and sovereign wealth funds can become "field stewards," shaping the flow and quality of capital in ways that stabilize both economies and ecosystems. They could stress-testing portfolios against climate shocks, decarbonizing balance sheets, and directing credit flows toward low-carbon sectors. For example, they could invest in climate resilience, next-generation biomaterials (such as mycelium and low-lignin wood), and carbon security indices that track systemic fragility alongside opportunity. For the Global South, sovereign climate funds offer a path to self-determined investment, reducing dependency and enabling leapfrogs to low-carbon infrastructure.

Risk and the New Logic of Hedging

In the quantum field model, risk is not eliminated—it is redistributed and transformed. Instruments like carbon futures, biodiversity credits, and parametric insurance create hedges that simultaneously incentivize regenerative action. The aim is not merely to protect portfolios but to embed resilience within the foundations of finance—capital systems that absorb shocks while reinforcing planetary stability.

The Emotional Field of Capital

If capital is a field, then trust, narrative, and emotion are part of its physics. Confidence can move markets faster than data; shared purpose can unlock flows that metrics alone cannot. Field-oriented finance must cultivate *financial ecosystems of care*, where capital strengthens long-term planetary habitability rather than extracting short-term gain. We must ask: *What does money represent? What values does it carry? Whom does it serve, and whom does it exclude?* To create a regenerative future, finance must be reframed as an act of care rather than conquest, prioritizing long-term planetary well-being over short-term profit.

What Comes Next

The next chapter will extend this "field" logic into governance—designing institutions as adaptive, polycentric networks that operate in resonance with the financial, ecological, and social fields around them. Chapter 12 will propose how institutions must be redesigned to operate as living, adaptive networks—leveraging polycentric leadership, participatory finance, and quantum feedback loops.

Example

Sovereign climate funds in the Global South* can invest in ecosystem services (e.g., mangroves, wetlands) with **measurable, multi-decade value**—ecological and economic.

Case Insight: A New Generation of Carbon Credit Cards

This quiet financial innovation is reshaping how individuals interact with the carbon economy. For instance, Sweden's **DO Black Card** not only tracks but actively limits a user's carbon spending, reframing climate action as a tangible financial boundary. Switzerland's **Climate Credit Card** couples each purchase with verified offsets, while the **Aspiration Zero Card** uses incentives like tree-planting to reward low-carbon consumption. West Bengal's proposed **Carbon Credit Card** appears to be one of the earliest attempts in India to connect individual low-carbon behavior with public-policy incentives through a credit-like mechanism. While widely reported as innovative, public reporting more clearly confirms the announcement of the initiative than its full operational rollout. Together, these tools turn invisible impacts into visible and actionable signals. As these systems scale, they will likely integrate with platforms like the **eCarbon Card**, which extends the concept into a comprehensive behavioral and trading framework, enabling households to monitor, reduce, and even exchange carbon allowances in real time.

"Carbon is not just an externality; it is a currency of future habitability." This framing, when combined with digital wallets, loyalty

*The focus on "sovereign climate funds in the Global South" reflects both necessity and opportunity. Many nations in the Global South face acute climate risks while lacking access to long-duration, patient capital for decarbonization and resilience. Establishing sovereign climate funds empowers these countries to define their own investment pathways, leapfrog fossil infrastructure, and shift climate finance from dependency toward agency—redefining development on their own terms.

programs, and ESG-linked savings products, marks the beginning of consumer-side climate finance—a bottom-up complement to institutional mechanisms.

Table 11.2

Name	Country	Feature	Mechanism	Implication
DO Black Card	Sweden	Carbon Spending Limit	Caps when personal CO_2 quota is exceeded	Introduces hard behavioral constraint
Climate Credit Card	Switzerland	CO_2 Footprint Calculator	Tracks emissions per transaction	Raises awareness, soft feedback
Aspiration Zero	USA	Plant a Tree' with every purchase	Offsetting vs. reduction approach	Raises awareness, soft feedback
Carbon Credit Card	West Bengal (India)	Reward points with low-carbon behaviors	Linked to a carbon credit card system	Public-sector led pilot

Part IV: Governance and Change

Chapter 12

Quantum Governance

From Institutional Alchemy to Polycentric Systems

"We cannot solve our problems with the same thinking we used when we created them."

— *Albert Einstein*

Introduction: Governance in the Age of Entanglement

The climate crisis is not merely an environmental failure but a profound governance failure. Our dominant institutions, forged during eras of industrial simplicity, were built to optimize, regulate, and control linear processes. They excel at issuing permits, enforcing compliance, and coordinating supply chains. Yet they falter when confronting nonlinear realities characterized by tipping points, cascading feedback loops, and emergent behaviors.

In a world defined by interdependence and volatility, traditional governance appears brittle. Ministries operate in isolation. Cities plan for heatwaves without aligning strategies for water or mobility. Policies are drafted faster than they can be adapted. What we need is not mere policy reform but a radical redesign of governance itself—a deeper transformation we call *institutional alchemy.*

The Limitations of Linear Institutions

Modern governance is still largely grounded in Newtonian logic. It assumes predictability—forecast, plan, and execute. It depends on control, regulating inputs to produce expected outcomes, and prizes efficiency, striving to do more with less at greater speed.

These principles work well in bounded and relatively stable

systems, such as taxation or logistics. But they break down in nonlinear, adaptive, and interdependent environments where even small interventions can trigger unpredictable effects. Climate, biodiversity, and social equity challenges demand responses beyond this linear toolkit. This mismatch between institutional design and systemic reality has created what complexity scholars refer to as a *governance deficit.*

Rethinking Governance as a Living System

What if we stopped treating institutions as machines to be optimized and instead understood them as living systems? This shift requires moving:

- from hierarchies to networks
- from command-and-control to feedback-and-adapt
- from a singular focus on outcomes to an embrace of process agility
- from simplistic problem-solving to collective sense-making

Governance, in this sense, becomes *metabolic*: capable of learning, evolving, and healing. Designing such governance architectures means creating institutions that mirror the complexity of the systems they are meant to steward.

The Limits of Hierarchy in a Nonlinear Crisis

Modern governance is rooted in an architecture of command and control, where ministries set policies, agencies enforce compliance, and citizens are expected to follow rules. This model functions reasonably well in linear systems with predictable cause-and-effect dynamics.

However, the climate crisis and broader sustainability challenges are neither linear nor predictable. They are emergent, distributed, interdependent, and rapidly changing. Top-down governance often breaks down under such nonlinear stress. Climate shocks outpace the slow cycles of bureaucratic response. Citizens disengage when decision-making is opaque and unaccountable. Cross-border issues like emissions, water, and biodiversity refuse to respect national boundaries. Accelerating technological change undermines legacy regulatory frameworks. To navigate these complexities, we must evolve from rigid, authority-based systems to adaptive ecosystems of governance.

From Newton to Nodal: Governance as a Network

A Newtonian approach assumes that central control is necessary to maintain coherence, that uniform policies will deliver uniform outcomes, and that any deviation represents failure. A quantum perspective, by contrast, recognizes that systems can self-organize under the right constraints, that decentralized agents can achieve alignment through effective feedback, and that diversity of response is not a weakness but a source of resilience. This shift from centralization to distributed *nodal governance* brings us to the concept of polycentricity.

Polycentricity: A Primer

Coined by Nobel laureate Elinor Ostrom, polycentric governance describes systems with multiple, overlapping centers of decision-making authority, operating at different scales—local, regional, national, and global—with autonomy, interaction, and mutual accountability. In a polycentric water management system, for example:

- A local water council might manage its aquifer.
- A regional body coordinates upstream and downstream rights.
- A national agency enforces water quality standards.
- Global conventions govern cross-border resource use.

No single actor dominates; instead, rules and solutions emerge from negotiation, experimentation, and iterative learning. Polycentricity is governance conceived as *ecology*, not machinery.

Quantum Analogies for Governance

Quantum principles illuminate this shift in governance architecture:

- **Superposition** – Multiple governance models can co-exist, each customized for its context, balancing competing priorities such as growth and equity without collapsing into false binaries.
- **Entanglement** – Local, national, and global policies are causally linked across both space and time; decisions in one arena can instantly shape outcomes elsewhere.
- **Observer Effect** – Measuring policy outcomes influences behavior, making metrics a powerful design tool.
- **Uncertainty Principle** – No policy can optimize for all variables simultaneously, requiring careful navigation of trade-offs.
- **Tunneling** – Innovation can bypass institutional inertia, with citizen labs and cross-sector collaborations creating new pathways.
- **Decoherence** – Systems fail when feedback loops are broken or when key actors are excluded from the decision-making web.

Real-World Polycentric Models

Polycentricity is not theoretical—it is already shaping governance worldwide.

- **Kerala's People's Planning Campaign (India)** – Beginning in the mid-1990s, nearly 40% of the state's budget was devolved to local bodies. Through village assemblies (*gram sabhas*), communities influenced priorities in housing, water, and sanitation—building resilience that proved critical during the 2018 floods.
- **Paris Agreement (2015)** – Encouraged nations to submit their own Nationally Determined Contributions (NDCs), balancing sovereignty with shared accountability, and using a "ratchet mechanism" for iterative ambition.
- **Guardianship Models (Ecuador, Canada)** – Ecuador's

constitution grants legal rights to nature; Canada's First Nations manage ecosystems through formal guardianship roles.

- **Amsterdam's Doughnut Economics (2020)** – Measures success by ensuring social foundations without breaching ecological ceilings, influencing zoning, procurement, and neighborhood design.
- **C40 Cities** – Municipalities across continents share climate data, policy tools, and best practices, often outpacing national governments.
- **Planetary Data Commons** – Platforms like Copernicus and Global Forest Watch make satellite data openly available for joint action.
- **Carbon Markets** – Interactions between private sector standards (e.g., Verra, Gold Standard) and national systems create distributed yet interoperable governance.

Governing for Emergence, Not Control

Linear systems ask *what rule should be applied*; complex systems ask *what enabling constraints allow beneficial patterns to emerge.*

Enabling constraints—rules, boundaries, or protocols—create coherence without imposing rigidity. A traffic roundabout limits speed and direction but improves flow. APIs in software development constrain interactions but enable massive interoperability. Ecological farming bans pesticides, which fosters biodiversity.

In governance, enabling constraints might include:

- shared climate targets across jurisdictions
- data interoperability protocols
- dedicated budgets for community co-design processes

Quantum governance emphasizes:

- protocols over rigid policies
- guardrails instead of micromanagement
- incentives over enforcement
- transparency to build trust

The eCarbon Card illustrates this: Who allocates allowances? How are data rights protected? Can allowances be traded? What safeguards exist for vulnerable populations? The answers determine whether innovation thrives or fails.

The Role of Digital Infrastructure

Digital infrastructure is the connective tissue of entangled governance. Distributed ledgers ensure transparency and traceability. Smart contracts automate compliance and feedback loops. Tokenized governance allows participatory decision-making. Digital identity platforms like the eCarbon Card ensures equitable access. AI analytics support real-time adaptation and scenario testing. These tools enable coordination without monopolistic control.

Ethics in Entangled Governance

Structure without ethics risks replicating existing inequities. Critical questions include:

- Who defines sustainability?
- Whose knowledge systems are valued?
- How are future generations represented?
- How is dissent handled?
- What rights are afforded to non-human entities?

Quantum governance acknowledges that truth is relational and justice participatory.

While this book has emphasized structures such as multi-domain entanglement and emergent adaptive governance, other models also capture the resonance of systemic change. The Quantum Change Spiral, outlined in **Appendix XV**, expands this lens by showing how transformation emerges simultaneously across fields of identity, perception, action, collaboration, and amplification.

From Technocracy to Quantum Democracy

The future lies in *coordinated decentralization*—where global frameworks set overarching goals while local actors innovate. Digital infrastructure enables continuous participation rather than episodic voting.

Design principles for quantum governance include:

- **Nestedness** – Linking local to global through interoperable protocols.
- **Reciprocity** – Two-way feedback between governed and governors.
- **Pluralism** – Multiple legal and cultural systems coexisting.
- **Reflexivity** – Institutions evolve their own logic through learning.
- **Coherence without uniformity** – Alignment of purpose without identical rules.
- **Ethical embedment** – Justice, care, and ecological boundaries at every level.

What Comes Next

The next two chapters explore carbon indices and economic signals as distributed nervous systems (Chapter 13), and decision-making under uncertainty (Chapter 14). Governance may be the scaffolding of systemic change, but the soul lies with people. The task ahead is to move from *institutions to individuals*—to become nodes of coherence and creativity in a nonlinear, interconnected world.

Chapter 13

Carbon Indices and Market Intelligence

Pricing the Invisible, Forecasting the Emergent, and Signaling Systemic Coherence

> Carbon is not just a molecule. It is a message—a signal through which the biosphere and economy speak to each other.

From Molecule to Market: The Rise of Carbon as Currency

Carbon was once a passive participant in economic life—extracted, burned, and discarded without consequence. In the Anthropocene, it has become the central organizing metric of planetary governance. Carbon is now:

- a molecule
- a risk
- a cost
- a target
- a tradable asset
- moral and strategic signal

Carbon pricing—through taxes, cap-and-trade systems, and shadow pricing—has become a cornerstone of climate policy. Yet these mechanisms often suffer from fragmentation, volatility, and political resistance. In a nonlinear world, the challenge is not just to price carbon, but to interpret the shifting patterns that determine its value. Markets are young, fragmented, and hypersensitive to both physical shocks and narrative swings.

What is needed now is a quantum leap in carbon market intelligence: tools that make carbon visible, comparable, and predictive, while reflecting systemic risk, social impacts, and planetary thresholds. This means creating a shared signal system—akin to a "carbon nervous system"—that helps governments, businesses, and citizens understand, coordinate and act upon the cost of emissions in real time.

This chapter explores how carbon indices, neural networks, and behavioral models can create a living, learning signal system—a carbon nervous system—for a decarbonizing world.

What Is a Carbon Index?

A carbon index is a composite benchmark that aggregates the prices of carbon allowances and offsets across markets—such as the EU ETS (Emission Trading System)[130], California's Cap-and-Trade Program[131], and the RGGI (Regional Greenhouse Gas Initiative)[132]—to track the evolving cost of emitting CO_2 equivalents. Like the S&P 500 for equities or the CPI for inflation, it acts as both a mirror and a compass—reflecting current conditions while guiding expectations.

From a quantum perspective, a carbon index is not just a price tracker. It is a field function—a synthetic signal integrating:

- market prices from multiple jurisdictions
- policy momentum
- technology deployment rates
- narrative sentiment
- risk-opportunity spreads

This multidimensional nature gives it behavioral leverage—shaping decisions through visibility and comparability, even without direct enforcement.

Why Do We Need Carbon Indices?

Carbon indices serve multiple functions: they create transparency, reveal market trends, and enable effective pricing of carbon-related risks. In a world of fragmented emissions trading systems, these indices provide coherence. Without them, volatility and uncertainty make long-term planning difficult for businesses and policymakers alike.

Volatility in carbon markets often stems from policy announcements, energy market shocks, macroeconomic fluctuations, and extreme weather events. Technological disruptions add another layer of complexity. For example, breakthroughs in Direct Air Capture (DAC) by companies like Climeworks and Carbon Engineering, now scaling toward million-ton capacity, can reshape the demand for carbon allowances by increasing the supply of removals, potentially dampening prices. Similarly, the rapid cost declines in green hydrogen—boosted by subsidies like the U.S. Inflation Reduction Act and the EU Green Deal—enable faster decarbonization in hard-to-abate sectors, altering projected allowance demand. Methane abatement technologies such as GHGSat's satellite monitoring platforms can rapidly cut short-lived pollutants, creating temporary price dips. Bio-based innovations, including low-lignin trees, mycelium composites, and kelp-based fibers, reduce emissions embedded in supply chains, while carbon-aware AI for grid optimization lowers operational emissions in the energy sector, all of which influence price dynamics.

The United States Carbon Index (USCA): A Case Study

Our research proposes the **United States Carbon Index (USCA)**[133]—a synthetic benchmark designed to integrate price signals from California's Cap-and-Trade Program (CARB), the Regional Greenhouse Gas Initiative (RGGI), and Washington State's Climate Commitment Act (CCA). It is calculated using monthly weighted averages of market prices, adjusted for coverage scope, liquidity, and compliance status, and is enhanced by a neural network model trained on macroeconomic indicators such as oil and gas prices, the S&P 500, the Federal Funds Rate, the Consumer Price Index, and industrial production data.

The USCA is significant because it forecasts six-month forward carbon cost trajectories, helping corporations and investors hedge exposure while guiding long-term abatement strategies. By providing a transparent and evidence-based signal, it encourages convergence among state-level policies and sends a clear message to national and global markets. More than a price, the USCA acts as a probabilistic pulse of the carbon economy—a dynamic indicator of systemic health and risk.

Neural Networks for Carbon Forecasting

Traditional linear models fail to capture the dynamics of carbon markets because these markets are shaped by nonlinear feedback loops, policy shocks, and complex economic interactions. Artificial neural networks (ANNs) provide a quantum-aligned alternative, capable of learning intricate patterns, handling multi-modal data (including prices, sentiment, weather, and policy changes), and adapting over time with new feedback. ANNs can output confidence intervals rather than single-point forecasts, making the carbon index predictive, interpretable, and responsive. In this way, markets shift from being perceived as chaotic noise to becoming sources of intelligence and foresight.

Behavioral Carbon Signals

Markets are not only shaped by assets but also by human agents with biases, heuristics, and social identities. Effective carbon indices must integrate behavioral economics, game theory, and cultural analytics to anticipate responses to price signals. When carbon prices rise, firms may invest in energy efficiency, shift procurement to low-carbon suppliers, accelerate R&D into alternative materials, or adopt internal carbon pricing mechanisms to future-proof operations. Others may seek offsets through nature-based or engineered removals, reevaluate high-carbon product lines, lobby regulators to shape future rules, form consortia for joint decarbonization infrastructure, or even exit high-emissions geographies.

These reactions highlight how carbon indices function not just as economic indicators but as levers for structural transformation across industries. By integrating behavioral insights, indices can inform complementary policy instruments—ensuring that market signals lead to systemic decarbonization rather than unintended distortions.

Democratizing Carbon Market Intelligence

Carbon indices must not remain the preserve of traders, investors, and policymakers; they must also serve citizens. Embedded in **eCarbon Cards**, for example, they can inform households about upcoming price trends, such as a projected increase in fuel costs, while linking individual choices to broader carbon goals. When connected to carbon dividend schemes, these indices can determine payout tiers, creating a direct financial feedback loop between personal behavior and planetary impact. Public dashboards and gamified neighborhood performance platforms can further democratize carbon literacy, transforming carbon awareness into a civic right rather than an expert privilege.

Ethical and Political Design Questions

The construction of carbon indices raises critical ethical questions. Who designs the index—private data firms, public institutions, or multistakeholder bodies? Which emissions are counted, and how are they measured? Are co-benefits like equity and biodiversity accounted for, or is the focus solely on CO_2? How are the voices and realities of developing countries represented in global benchmarks? A quantum carbon index must be pluralistic, transparent, and reflexive, embedding ethics and justice into its very code.

Polycentric governance can distribute trust:

- exchanges supply raw market data
- institutes harmonize methodologies
- civic tech translates complexity into accessible public tools

Ethics, equity, and transparency must be embedded in the index's DNA.

Integration with Global Systems

Carbon indices can do more than track prices; they can anchor **climate-linked currencies** like a hypothetical "CarbonCoin" tied to verified offsets, drive Scope 3 transparency in corporate supply chains, inform climate risk disclosures under frameworks like Task Force on Climate-Related Financial Disclosures (TCFD) and International Financial Reporting Standards (IFRS), and enable agile policies such as dynamic carbon taxes or responsive subsidies. When integrated with digital twins of cities, industries, or energy systems, carbon indices become part of a planetary feedback loop—a real-time intelligence layer for collective climate action.

From Signal to Synchrony

Markets often reveal fragmentation and misaligned incentives, but a quantum carbon index can foster synchrony—aligning prices with planetary thresholds, connecting individual behavior to systemic outcomes, and linking local actions with global consequences. By combining the power of AI, IoT, and blockchain, these indices can become choreography platforms for climate action, harmonizing diverse efforts into a coherent, collective movement.

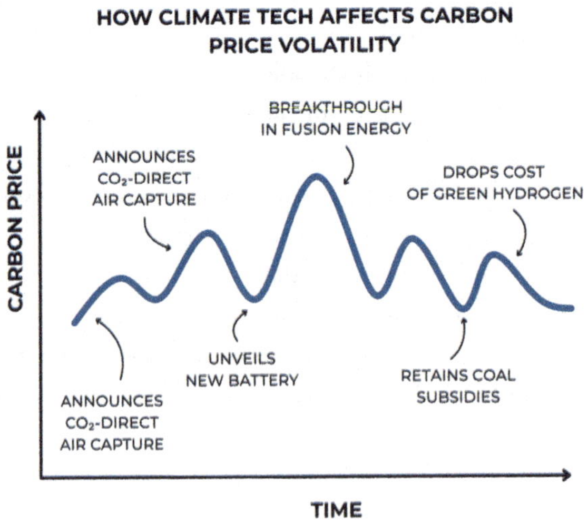

Figure 13.1: Architecture of a Quantum-Aware Carbon Index

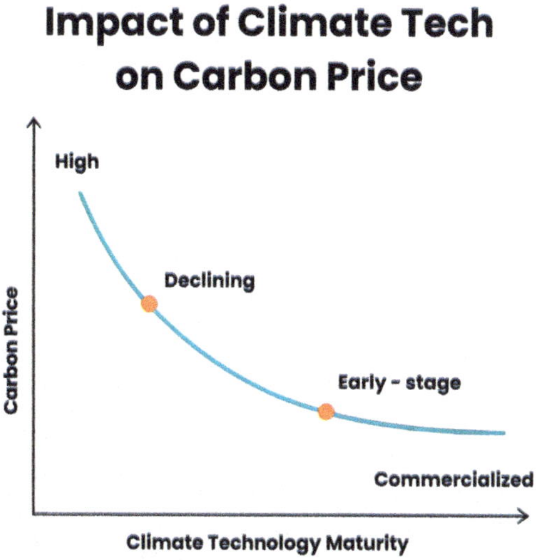

Figure 13.2: Feedback Loops in Market Intelligence

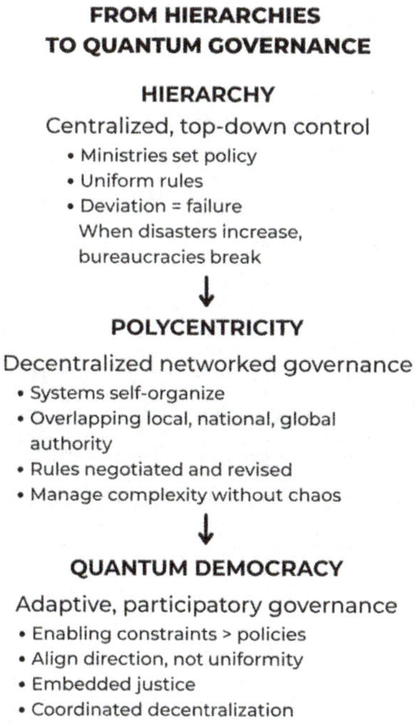

Figure 13.3: Policy, Technology, and Narrative Drivers of Carbon Price Volatility

Table 13.1 – Carbon Index Tools vs. Behavioral Signals

Dimension	Carbon Index Tools (Technical Levers)	Behavioral Signals (Human Levers)
Purpose	Aggregate and forecast carbon prices across markets	Translate prices into personal and social awareness
Core Mechanism	Data integration, AI modeling, and market benchmarking	Nudges, social norms, gamification, and feedback
Examples	USCA, EU ETS benchmarks, neural network forecasting	eCarbon Card feedback, peer comparison dashboards
Target Audience	Policymakers, corporations, investors	Households, communities, consumers
Outcome	Market intelligence, risk hedging, policy alignment	Behavior change, carbon literacy, cultural shifts

Closing Thoughts, What Comes Next

Carbon indices, if designed with complexity in mind, can be more than market trackers—they can be instruments of coordination, trust, and acceleration. In a nonlinear climate crisis, the most valuable intelligence is not the data we have but the questions it prompts and the agility it enables.

As we move forward, we shift from instruments to intention—from systemic designs like carbon indices to the deeper challenge of acting upon these signals in conditions of uncertainty. The next chapter explores decision-making when the future is entangled, outcomes are probabilistic, and leadership must navigate within a quantum field of possibility.

Chapter 14

Time, Uncertainty, and Climate Leadership

Making Decisions When the Future Is Entangled and the Clock Is Irreversible

> The paradox of climate leadership is this: you must act with urgency while embracing uncertainty. The clock does not wait for clarity.

The Leadership Dilemma in a Nonlinear Era

Leadership is often framed as the art of prediction, persuasion, and planning: analyze the data, model the future, make the optimal choice, and convince others to follow. Yet the climate crisis defies this model. It is nonlinear, deeply uncertain, moral as well as material, rapid and slow at once, global in scope yet intensely personal in impact. Traditional leadership—anchored in certainty and linear logic—crumbles in this domain. What is needed is quantum leadership, which operates across multiple timescales, integrates paradoxes, acts with incomplete information, holds space for emergence, and invites participation rather than control. This chapter explores how to lead when certainty is a luxury, ambiguity is structural, and delay becomes betrayal.

Why Climate Decisions Are Uniquely Difficult

Climate change challenges leadership in ways that are profound and multidimensional. At its core lies **deep uncertainty**: we cannot fully predict how fast tipping points will unfold, how societies will adapt, or which technologies will scale and when. Rather than a single trajectory, there exists a probability field of multiple, often conflicting futures.

Another complexity is the **time lag** between actions and outcomes. Emissions released today lock in consequences—such as sea-level rise—decades later, while trees planted now may not deliver their full sequestration potential until 2040 or beyond. Traditional forecasting struggles to cope with such delayed feedback, compounded by the systemic complexity of climate and sustainability systems. Policies such as carbon taxes or emissions standards often take years to influence market behavior, technology transitions, or emissions outcomes, creating a gap between action and visible results. Green infrastructure investments—mass transit systems, renewable energy grids—may take a decade or more from planning to completion, with emissions reductions accruing gradually. Even the global carbon budget is path-dependent, with today's cumulative emissions committing us to tomorrow's warming.

Ecological feedbacks, such as permafrost thawing or ocean acidification, unfold too slowly to trigger immediate political urgency, while financial markets and supply chains often resist rapid pivots due to sunk costs, regulatory inertia, and long-term contracts. Climate data itself may take years to cycle through scientific models, policy debates, and implementation, by which time conditions may have shifted. These time lags create a dangerous blindspot, allowing systems to drift toward critical thresholds long before the consequences are apparent, resulting in delayed and insufficient responses.

Beyond uncertainty and time lag, leaders face *cognitive overload.* The climate crisis is abstract and sprawling, while human brains are wired for immediacy rather than slow-moving complexity. Moral ambiguity compounds this challenge: should priority go to mitigation or adaptation, growth or equity, global risk or local livelihoods? Leadership here is not just strategy—it is ethical navigation. Add to this decision fatigue from overlapping crises—pandemics, conflicts, economic shocks—and it becomes clear that climate leadership must resist distraction and despair while staying focused on long-term transformation.

Quantum Decision-Making: A New Model

In quantum systems, outcomes are probabilistic, states evolve through entanglement, observation changes the system, paths interfere with each other, and time can flow in nonlinear ways—sometimes reversible, sometimes compressed. These principles offer valuable analogies for leadership under climate uncertainty. Quantum decision-making does not seek a single correct answer; it creates conditions where multiple possibilities can be explored, tested, and adapted. It embraces uncertainty as a strategic asset rather than a threat, allowing leaders to act while learning and co-evolving with the system.

The Six Pillars of Quantum Leadership

The first pillar is to **act before you're ready**. In a deterministic mindset, decisions wait for perfect data; in quantum leadership, action itself shapes the future. This means piloting regenerative agriculture even before the full business case is proven, launching clean energy transitions while fossil fuel subsidies are still being dismantled, and crafting narratives that lead public discourse even before consensus forms. As the saying goes: "Leap, and the net may emerge."

The second pillar is to **balance fidelity and flexibility**. Quantum systems require coherence and superposition; leaders must anchor themselves in unshakable values while adjusting their methods fluidly as conditions evolve. A company might hold fast to its net-zero goals while remaining open to changing its technological pathways as new innovations arise.

The third pillar is to **navigate with scenario constellations** rather than linear forecasts. Traditional planning models rely on a single "business-as-usual" future, leaving organizations blind to emergent shocks or tipping points. Quantum leadership explores **multiple plausible futures**—narratives grounded in real drivers of change—to test resilience and stretch strategic thinking. This approach embraces uncertainty, surfaces weak signals, enhances resilience, promotes creativity, and reduces overconfidence. For example, a coastal city may examine futures shaped by rapid ice melt, a green technology revolution, stagnant global action, or climate-driven migration, ensuring that infrastructure decisions are held under most scenarios. A national utility might explore combinations of climate regulations, storage breakthroughs, geopolitical tensions, and demand shifts, staging investments that remain viable across all. In the textiles and nonwovens industry, companies can model futures where bio-based feedstocks dominate versus scenarios where synthetic circularity or carbon taxation accelerates recycling, designing product lines with modularity and reversibility. Scenario constellations highlight system entanglement, allow for simultaneous consideration of multiple pathways, and embrace nonlinearity and tipping points, turning uncertainty into a source of strategic strength, and thus reinforce as summarized below:

- **Entanglement** – Reveals interdependence of systems and actors across futures.
- **Superposition** – Allows simultaneous consideration of multiple pathways without collapse.
- **Uncertainty** – Makes it a strength, not a threat.
- **Observer Effect** – Encourages reflection on how present assumptions shape future readiness.
- **Decentralization** – Allows for distributed scenario development and localized responses.
- **Nonlinearity** – Embraces tipping points, feedbacks, and emergent discontinuities.

This approach is especially vital for sustainability, where change is nonlinear and path-dependent. By navigating through plausible futures rather than anchoring to one, organizations become adaptive stewards rather than reactive responders.

The fourth pillar is to lead with emotional intelligence. Uncertainty breeds fear, grief, and inertia. Quantum leaders name these emotions, normalize ambiguity, and inspire through meaning rather than metrics. They do not pretend to have certainty; they cultivate trust and confidence even in its absence.

The fifth pillar is to share the decision space. In complex systems, no single leader sees the whole picture. True leadership invites participatory governance, bringing citizens into deliberation, scientists into policymaking, youth into priority-setting, and artists into storytelling. This distributed, inclusive model reflects the entanglement of knowledge and values needed to navigate a turbulent world.

The sixth pillar is to design for adaptive iteration. Quantum leaders present strategies not as fixed plans but as prototypes, continuously tested, scaled, or pivoted based on real-time learning. Feedback loops—such as dashboard indicators, citizen reporting, and live emissions or ecosystem health metrics—inform iterative adaptation. Governance, in this framing, becomes choreography rather than command.

The Role of Ritual, Pause, and Reflection

In classical leadership models, more data is equated with better decisions. In quantum leadership, the ability to pause becomes a source of power. Pausing allows leaders to reflect on unintended consequences, revisit values, celebrate small victories, and create rituals of care for teams and communities. This cultivates temporal coherence—the ability to act with alignment across past, present, and future.

The Paradox of Fast Patience

Climate leadership demands both urgency and patience. It requires acting swiftly to meet planetary boundaries while allowing the slower rhythms of social, behavioral, and ecological transitions to unfold. Quantum leaders master this polarity by moving quickly on low-regret actions, such as phasing out fossil subsidies, while advancing social transformations at a more deliberate pace. They learn fast, scale slow, and combine speed with humility and courage.

Leadership as Pattern Recognition

Quantum systems do not hand down commands; they offer patterns. A sudden uptick in village solar adoption, a surge in youth climate activism, a hidden opportunity in waste-to-value chains, or a tipping point in urban electric transport—all are subtle signals that require leaders to connect the dots and trust in emergent possibilities. This is not reactive management but pattern literacy—a way of seeing complexity as a tapestry of interlinked signals.

Courage as Entangled Will

Quantum climate leadership is less about certainty and more about holding the weight of the unknown while acting anyway. It invites others into the field of possibility and accepts the consequences without collapsing into paralysis. Courage, as C.S. Lewis observed, is the form of every virtue at the testing point—and we are now at that testing point. Radical acceptance does not mean resignation; it is the acknowledgment of climate consequences already in motion—melting glaciers, rising seas, shifting ecosystems—not as failures but as part of a turbulent transformation that still offers agency. To accept without paralysis is to grieve losses while continuing to build regenerative solutions, to acknowledge locked-in emissions while accelerating adaptation and innovation. Paralysis comes from seeing consequences as binary: either we save the planet, or we lose it. In reality, complex systems provide multiple inflection points. Even irreversible changes can be buffered, delayed, or adapted to, and acceptance becomes the grounding force from which moral clarity and adaptive action arise.

Part V: Designing the Quantum Leap

Chapter 15

Designing for Emergence

In complex adaptive systems, change rarely follows a linear blueprint. It unfolds through interactions—between people, policies, technologies, and ecosystems—that cannot be fully predicted in advance. The most resilient solutions are not those that aim to control outcomes but those that create the conditions for emergence: the spontaneous development of new structures, behaviors, and possibilities from the interplay of system components.

Emergence is not an accident. While we cannot dictate exactly what will emerge, we can design enabling conditions, feedback pathways, and adaptive capacities that allow the "unexpected" to be constructive rather than destructive.

From Control to Choreography

In a Newtonian mindset, a designer sets a plan, enforces rules, and measures success against pre-set milestones. In an emergent mindset, the role shifts from controller to choreographer—guiding interactions, setting boundary conditions, and adjusting in real time as patterns form.

Example

The transition to regenerative agriculture in Costa Rica did not follow a rigid script. Instead, networks of farmers, NGOs, and government agencies co-created practices over time, experimenting with soil amendments, crop rotations, and biodiversity corridors. The outcome—higher yields, improved soil health, and rural incomes—emerged from decentralized innovation supported by shared learning hubs.

Feedback-Enabled Learning Ecosystems

Emergence thrives on feedback. Without the ability to sense, interpret, and respond to changing conditions, systems risk getting locked into maladaptive paths.

Design Principles

- **Short Feedback Loops** – Move from annual reviews to weekly or real-time sensing (e.g., IoT sensors for soil moisture or grid loads).
- **Bidirectional Flow** – Feedback should travel not only from field to headquarters but also from communities to policymakers, and back again.
- **Plural Metrics** – Combine quantitative and qualitative signals—CO_2 levels, biodiversity counts, and local cultural well-being indicators.

Case

In Copenhagen's climate adaptation strategy, flood risk mapping is updated in real time with citizen-reported data, enabling micro-adjustments to urban drainage systems before disasters occur.

Adaptive Innovation Portfolios

A system designed for emergence must cultivate diversity—not just in technology, but in approaches, actors, and scales.

Portfolio Design

- **Parallel Experimentation** – Fund multiple solutions simultaneously (solar microgrids, demand-response programs, peer-to-peer energy trading).
- **Stage-Gating with Flexibility** – Keep high-potential projects alive even if early results are ambiguous—emergent value may surface later.
- **Cross-Pollination** – Create spaces where teams from different domains (energy, mobility, agriculture) exchange learnings.

Analogy

Think of an innovation portfolio as a coral reef—multiple species coexisting, some dominating for a while, others taking over when conditions change.

Complexity as Opportunity

Most organizations see complexity as a threat to be simplified. In reality, complexity offers more surface area for innovation. Interdependencies create more entry points for change—if we know how to find and activate them.

Opportunity Mapping Tool

- **Map Entanglements** – Identify cross-sector linkages (e.g., how waste heat from industry could power greenhouses).
- **Spot Positive Spillovers** – Where might one intervention unlock multiple co-benefits?
- **Design for Modularity** – Ensure components can evolve independently without destabilizing the whole system.

Guardrails for Healthy Emergence

Emergence without constraints can lead to instability or inequity. "Enabling constraints" focus energy and resources toward shared goals while leaving room for adaptation.

Examples of Enabling Constraints

- minimum biodiversity thresholds in reforestation programs;
- open data protocols for all publicly funded smart-city projects;
- carbon intensity caps for energy trading platforms.

Emergence in Practice: Three Case Snapshots

1. **Kenya's M-Pesa and Solar Pay-as-You-Go:** Mobile money infrastructure enabled emergent energy access models without central planning.

2. **Amsterdam Circular Economy Agenda:** City set broad goals (50% circular by 2030) but left space for neighborhood-led pilots, leading to unexpected upcycling ventures.
3. **California's Demand Flex Markets:** Policy frameworks allowed for multiple demand-response providers to emerge organically, increasing grid flexibility.

Designing for Emergence: A Practical Checklist

- **Boundary Conditions** – Have we defined the "non-negotiables" for equity, ecology, and resilience?
- **Portfolio Diversity** – Are we running multiple approaches in parallel?
- **Feedback Channels** – Can we detect change quickly and act on it?
- **Learning Infrastructure** – Do we have spaces and rituals for sharing failures and insights?
- **Scalability Pathways** – Can successful local patterns scale without losing their adaptability?
- **Exit and Evolution Plans** – Can we gracefully retire interventions that no longer serve the system?

Closing Thought

Emergence is not magic. It is the product of intentional design, openness to surprise, and a commitment to continuous learning. In the face of a nonlinear crisis, we cannot blueprint the future—but we can build the fertile ground from which better futures can grow.

The quantum walk reminds us that progress is less about predicting where we will land and more about multiplying the possibilities for where we could arrive. It is about designing conditions where regenerative agriculture, decentralized energy, or circular economies can emerge not as isolated projects, but as entangled patterns of resilience.

To design for emergence is to embrace humility: knowing we cannot control outcomes, yet also courage: stepping forward to shape boundary conditions, feedback loops, and enabling constraints. It is to recognize that in complexity, every interaction matters—a village solar cooperative, a new carbon index, an Indigenous stewardship practice, or a change in consumer behavior can all ripple outward, synchronizing into systemic transformation.

The closing lesson, then, is coherence. When values, institutions, and innovations resonate across scales, what emerges is not chaos but possibility: a living choreography of sustainability that adapts, self-corrects, and grows more resilient with time. This is the true work of quantum leadership: not to command systems, but to cultivate the fields where emergence becomes inevitable.

Transition to the Collective

The next chapter shifts the lens from the individual to the collective. What does it mean for citizens, communities, networks, and societies to become quantum actors? We will explore emerging models of social identity, climate citizenship, collective intelligence, and cultural rituals of transformation. With Chapter 16, we move toward the idea of the quantum citizen—our final step before integrating all paths into a future-back design

Chapter 16

The Role of the Quantum Citizen

Entangled Agency in the Anthropocene

> A quantum citizen doesn't ask, 'What can I do?' They ask, 'What can I ripple?'

From Spectator to Participant

In the prevailing narrative of climate action, individuals are often trapped in a paradoxical frame: either they are cast as consumers who must "buy greener products," or they are told that true change is beyond their reach, dependent on policies and industries. This duality breeds confusion, fatigue, and detachment. A quantum worldview, however, recognizes that systems are deeply entangled and feedbacks are powerful. Individuals are not passive endpoints of systemic forces but coherent nodes in a distributed and dynamic field of influence. A quantum citizen does not merely react to the climate crisis but participates in shaping its unfolding reality through intentional action, alignment with values, narrative building, and collective engagement. The individual is not the smallest unit of change but the smallest scale at which larger wholes can emerge.

Agency in the Age of Systems

Traditional civic roles—voting, donating, protesting—remain essential, but they are limited by their episodic nature and the scale of impact they can achieve alone. Quantum citizenship is multimodal, weaving personal ethics with systems awareness, linking private choices with public influence, and moving fluidly across domains such as energy, food, finance, and information. What defines a quantum citizen is not the magnitude of their actions but the coherence between values, knowledge, behavior, and context. This coherence generates ripple effects across networks by shifting social norms, signaling preferences to markets, role modeling cultural values, and influencing governance from the ground up.

From Carbon Footprint to Carbon Imprint*

For years, climate communication focused on reducing the "carbon footprint," urging individuals to eat less meat, fly less, and adopt energy-efficient appliances. While these actions remain valuable, the footprint framing is inherently limiting. It places the burden of systemic failures on individuals, overlooks the structural and institutional barriers that shape behavior, and often induces guilt rather than empowerment. A richer metaphor is the "carbon imprint," which captures the totality of one's influence—how choices ripple through supply chains, signal new market norms, and shape institutional behaviors. The imprint is not just about emissions avoided; it is about the narratives one normalizes, the behaviors one validates, and the institutions one chooses to challenge or uphold. It moves the conversation from subtraction to resonance, from personal restraint to systemic influence.

While the carbon footprint is static and narrow, the carbon imprint is relational and dynamic. For instance, adopting a plant-based diet not only lowers direct emissions but also signals demand to food companies, reduces deforestation pressures by curbing livestock production, and inspires similar shifts in a social circle. The carbon imprint measures not just what we emit but what we enable, accelerate, or disrupt across the systems we touch.

* "Footprint" is about subtraction. "Imprint" is about signal. A quantum citizen leaves a resonant signature on the social field.

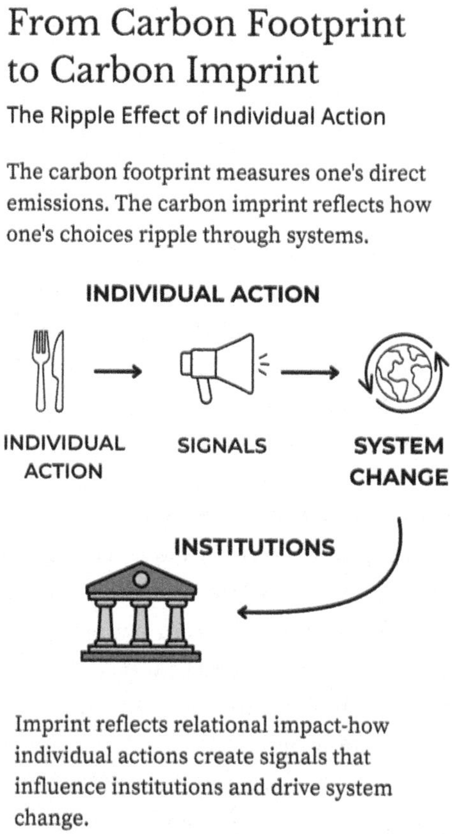

Figure 16.1

The Five Spheres of Quantum Citizenship

Quantum citizens exert influence across five interconnected spheres, not through sheer scale but through distributed coherence. First, they practice embodied habits that form powerful feedback loops. Plant-based eating, refusing fast fashion, or commuting by bike may seem like individual acts, but they signal new norms that others observe, absorb, and emulate, creating local rituals with global echoes.

Second, they activate narratives. We live inside stories, such as "green means sacrifice" or "the economy and ecology are at odds," which shape our collective behaviors. Quantum citizens challenge and rewrite these stories through blogs, podcasts, art, workplace conversations, and casual exchanges around dinner tables. They replace narratives of helplessness with ones of entanglement and agency, recognizing that a single voice can become an actuator of change.

Third, they align resources—time, money, and attention—with their values. Whether banking with ethical institutions, investing in climate funds or community solar projects, supporting mutual aid networks, or volunteering skills for transition initiatives, these choices represent capital flows as field functions rather than mere economic transactions. Even small re-allocations can trigger cascading signals through markets and supply chains.

Fourth, they circulate knowledge. Quantum citizens see themselves as nodes in a broader web of understanding. They translate complex science into accessible stories, teach systems thinking to colleagues or children, debunk misinformation, and curate actionable insights for their communities. Learning, in this frame, is not passive accumulation but a form of constructive interference—shaping what others can imagine.

Finally, they weave communities. Individual action gains power when linked to collective coherence. Starting a climate-focused book club, building neighborhood resilience hubs, facilitating town hall conversations, or creating local dashboards and art installations are all forms of social mycelium—dense, cooperative, and intelligent networks that grow strength through connection. Quantum citizenship, unlike mass-produced campaigns, scales like a living network rather than a product.

Quantum Citizenship Is Not Perfection—It Is Participation

Quantum citizens are not defined by environmental purity but by curiosity, commitment, and courage. They embrace contradictions without losing coherence: one can fly to attend a funeral and still advocate for better train subsidies, enjoy an occasional meat dish while supporting plant-forward policies, or work in a legacy industry while pushing for its transformation. The goal is not perfection but steady progress toward reducing dissonance and amplifying impact over time.

From Identity to Role

The question is no longer "Am I a climate person?" but "What is my role in this entangled moment?" Quantum citizens assume diverse archetypes—signal amplifiers, caregivers, provocateurs, builders, bridges, or protectors—each contributing uniquely to planetary stewardship. One need not do everything; the imperative is to do one's part, well, and in alignment with others.

Cultural Technologies of Change

Supporting quantum citizenship requires new cultural technologies—tools that make complex participation intuitive, joyful, and collective. Examples include eCarbon Cards for household-level carbon tracking, digital carbon wallets and local climate currencies, and "climate action time banks" where people exchange efforts rather than money. Rituals of grief, hope, and recommitment, alongside gamified story-worlds and community dashboards, make the invisible visible and transform climate action from an obligation into a shared cultural practice.

Barriers and Breakthroughs

Quantum citizens face barriers such as climate fatigue, institutional distrust, the overwhelming complexity of global crises, and polarization amplified by digital echo chambers. Yet breakthroughs are equally present: peer influence and shifting norms, local victories such as community-owned renewable energy, emotional literacy and collective care, regenerative storytelling and artistic expression, and youth-led visioning paired with intergenerational partnerships. A quantum citizen acknowledges barriers but flows around them, focusing on where ripples can create new currents of change.

The Quantum Citizenship Oath

"I accept that I am not separate from the systems I seek to change. I pledge to act with humility, coherence, and imagination. I will listen deeply, learn continuously, and participate generously. I am not alone. My signals ripple. My presence matters. I walk the future back through how I live today."

From Citizen to Field: What's Next

As we approach the final chapters, we begin to integrate the full spectrum of tools—systems thinking, finance, governance, carbon signals, decision-making, and citizen agency—into a coherent blueprint for transformation. In the next chapter, we begin to **synthesize and integrate**—explore how to walk the future backwards, assembling the actors, pathways, and instruments needed to design a regenerative world. Chapter 17, *Walking the Future Backwards*, marks the closing arc of this quantum journey through sustainability.

Chapter 17

Walking the Future Backwards

From Vision to Pattern, From Possibility to Probability

> In a quantum universe, the future is not a destination. It is a field. We walk toward it by walking from it.

The Trap of Forward-Only Thinking

The dominant planning paradigm in sustainability remains linear and forward-facing: set a 2050 net-zero target, extrapolate current trends, project innovation curves, and build policy models or backcasts based on assumed continuity. While useful in certain contexts, this approach often overestimates our predictive power and underestimates the potential of emergence. It assumes that the future will simply be an extension of the past—just cleaner, greener, and more efficient. But the Anthropocene doesn't conform to linear rules. Ecological thresholds are non-negotiable, technological disruptions unfold in nonlinear bursts, social tipping points defy modeling, and values or consciousness evolve in unexpected ways. We cannot merely forecast our way into a sustainable future. Instead, we must design from the future, treating it as a gravitational attractor and walking backward from it into today's strategies.

This chapter explores backcasting as a quantum design method, a way of working backward from desired futures to create actionable steps that bring those futures into view.

Backcasting vs. Forecasting

Forecasting begins with the present, extrapolating trends forward and assuming continuity, while backcasting starts with a future vision and works backward, embracing discontinuity and transformation. Forecasting is driven largely by data and modeling, seeking to optimize known variables, but backcasting is rooted in values, imagination, and systemic logic. It explores the reconfiguration of entire systems rather than incremental improvements and challenges existing institutional assumptions instead of reinforcing current power structures. Backcasting is not unrealistic or fanciful; rather, it anchors realism in desired possibility, allowing strategy to arise from what we want, not just from default momentum.

Quantum Time and the Ethics of Future-Making

In quantum theory, time is not absolute but relational, and events are shaped by both past and future boundary conditions. Our present choices do not merely follow from history; they also determine which future probabilities solidify into reality. Quantum sustainability thus asks difficult but vital questions: What futures are we loyal to? Which seeds must we plant today to make those futures more probable? And what risks must we take now—not because they are guaranteed to succeed, but because they are ethically required? This mindset is not naive optimism; it is strategic hope, grounded in courage and deliberate intent.

Building from Future Archetypes

To make this concrete, we can explore three future archetypes and walk backward from them into today's necessary actions.

The **Regenerative Civilization of 2070** is one in which planetary temperatures stabilize below 1.6°C, agroecological food systems dominate, energy is fully renewable, distributed, and democratized, and circular materials and nature-based industries thrive.

Governance is polycentric and participatory, and cultural identity is rooted in stewardship rather than extraction. To reach this vision, we must ask what technologies had to be deployed at scale by 2030, which value shifts enabled degrowth in the Global North, what financial architectures (such as post-EBITDA metrics) were normalized, and what role Indigenous knowledge and decentralized governance played in catalyzing the transition.

The **Cascading Crisis World of 2070** depicts a far darker future: global warming exceeding 2.4°C, frequent ecological and economic disruptions, mass climate migration, authoritarian backlashes, and biodiversity collapse beyond critical tipping points. Culture in such a world becomes dominated by nostalgia, fear, and reactive adaptation. Walking backward from this scenario, we must consider which warning signs were ignored in the 2020s, where political polarization obstructed systemic change, how financial capital failed to pivot, and what narratives collapsed public imagination and collective willpower.

The **Quantum Society of 2070** represents a middle but transformative path, where climate stabilizes below 2.0°C through unexpected means, and society reorganizes around relational values. Leadership becomes distributed, while behavioral tipping points are driven by youth, faith, art, and care. Digital, biological, and ecological intelligences co-evolve, and the economy is redefined as the stewardship of flows rather than the extraction of stock. Walking backward from this vision prompts questions about what new metrics, rituals, and platforms were invented, which cultural tipping points reshaped identity and aspiration, how AI and systems tools were harnessed to design coherence, and what forms of planetary citizenship emerged.

Tools for Walking the Future Backwards

Backcasting is operationalized through several powerful tools. Future scenario mapping uses multiple archetypes to stress-test strategies, surface blind spots, and expand the space of possibility. Systems backcasting takes a desired future—say, an 80% circular economy—and reverse-engineers the necessary technology deployment, policy frameworks, cultural shifts, and business models to achieve it, identifying leverage points in today's systems. Policy time machines immerse stakeholders in future conditions using augmented reality decision labs, participatory simulation games, or scenario-based citizen assemblies, allowing people to inhabit possible futures before they arrive. Finally, iterative prototyping replaces rigid plans with adaptive pilots, policy sandboxes, experimental governance models such as climate assemblies, and short feedback loops powered by real-time data. Rather than waiting for perfect solutions, this approach embraces the principle of failing informatively and adapting quickly.

The Role of Visionary Infrastructure

Infrastructure is the silent architecture of future possibility. To make regenerative futures more likely, we must embed future coherence into the design of our cities, technologies, and economic systems. This includes biophilic and climate-resilient urban design, digital systems that are open, ethical, and interoperable, localized and just food systems, decentralized and democratized energy grids, education built around systems thinking and critical foresight, and capital markets governed by entangled metrics of value rather than single-dimensional profit. These are not just technical fixes; they are the codebases of emergence, invisible scaffolds that shape visible futures.

Walking with Others

Walking the future back is not an individual exercise but a collective ritual of imagination and design. It requires the wisdom of elders who understand long-term shifts, the creativity of youth who can imagine beyond inherited constraints, the vision of artists to visualize new worlds, the rigor of engineers to ground ideas in feasibility, and the voice of citizens to root solutions in place, justice, and belonging. Such collective foresight builds not just policy options but political will and shared meaning.

The Ethics of Choosing a Future

Walking from the future requires deep moral reflection. Which future do we most want? Which do we most fear? Which do we most believe in, and which are we willing to serve even if we may not live to see it? These questions are not technical but ontological—they define the moral horizon we aim toward. The future is not a spreadsheet of probabilities; it is a tapestry of values, commitments, and shared imagination.

The Role of the Quantum Walker

A quantum walker holds multiple futures simultaneously, tuning present action to resonate with preferred futures while accepting that small, coherent actions ripple outward with great power. They understand that the path is made by walking, but that the walking begins from the future—moving backward with intention and alignment.

Transitioning to Synthesis

This chapter brings together systems, citizens, and time into a visioning practice grounded in both planetary realism and strategic imagination. In the final **Part VI: Putting It All Together**, we will integrate all the quantum principles explored in this book, proposing a new model for sustainability, leadership, and societal transformation. We will move from walking the future back... to living it forward. Let us now proceed to **Conclusion: A Quantum Theory of Change.**

Part VI: Putting It All Together

Afterword

A Quantum Theory of Change

Becoming the Coherence You Wish to See in the World

> Classical change begins with intention. Quantum change begins with entanglement. Once you realize you are part of the system, change is not a choice—it is a consequence.

Revisiting the Journey

We began this book with a provocative premise: the climate crisis is not just a linear problem to be solved with linear tools. Our economic models, governance systems, policy instruments, and cultural narratives are still anchored in a Newtonian worldview—one that assumes cause and effect are separable, that change is incremental, and that the whole is nothing more than the sum of its parts. Yet, climate, sustainability, and planetary futures unfold in ways that reflect quantum-like complexity—marked by nonlinearity, emergence, entanglement, feedback loops, observer effects, and pervasive uncertainty.

Through this lens, we embarked on a quantum walk across the key domains of sustainability—systems, signals, leadership, citizenship, time, and imagination—building toward a new framework for transformative action. In this final chapter, these insights coalesce into what can be described as a Quantum Theory of Sustainability Action: a model of change that is relational, adaptive, and deeply coherent.

The Limits of Classical Change Models

Most mainstream change models for sustainability rely on linear causality, predictable policy levers, and straightforward behavioral nudges. They are structured around binary dynamics such as top-down versus bottom-up, or deterministic narratives where technological progress is assumed to be the prime driver of transition. While these approaches can yield useful outcomes, they often fail to account for tipping points and systemic feedback, lack mechanisms to scale moral imagination, and remain blind to emergent behaviors that arise from complex, interconnected systems. Moreover, they rarely embed justice, care, or cultural meaning into their metrics of success. These models assume the world is stable and knowable—when in fact, the world we inhabit is dynamic, interdependent, and perpetually unfinished. A quantum theory of change acknowledges this complexity and seeks to harness it rather than control it.

Anchoring in Quantum Structures

Throughout the book, we have engaged with several paradigms that capture this nonlinear reality. **Multi-domain entanglement** illustrates how energy, food, water, carbon, and justice are bound together; progress in one is inseparable from progress in the others. **Constructive interference** shows that when aligned, signals from policy, markets, and culture amplify each other into waves of transformation rather than canceling out. **Observer responsiveness** reminds us that the very act of measuring and narrating sustainability changes the trajectory—metrics and stories are not neutral but catalytic. And **emergent adaptability** underscores that resilient systems are not rigid but evolve through iteration, feedback, and learning.

These structures have immediate practical benefits: they help policymakers design reinforcing loops, guide businesses toward adaptive innovation, and empower communities to find coherence across diverse local actions. They are not abstractions but usable lenses through which systemic change can be envisioned and enacted.

From Systems Thinking to Systems Embodiment

The shift we need is not only intellectual but embodied. We must evolve from simply thinking about systems to becoming living examples of those systems. Eating becomes a planetary act, banking becomes a moral statement, storytelling becomes infrastructural, governance becomes relational, and work becomes ecological. To "be the system you wish to reform" means internalizing the interconnections we advocate for and allowing them to guide our actions. This is the praxis of the quantum agent—not to fix systems from the outside but to transform them from within, through alignment, presence, and intentional design.

Quantum Leverage Points: A Reframing of Meadows

Building upon Donella Meadows' leverage points, a quantum perspective reframes the highest-impact interventions as possibility portals. Shifting the underlying paradigm—from extractive logic to regenerative thinking—is the most transformative move. Designing rapid, trusted, and visible feedback loops ensures systems can adapt and learn continuously. Narrative infrastructure becomes equally critical, as stories shape collective imagination and institutional action. Increasing relational density—deep trust and cooperation across silos—strengthens resilience. Developing temporal literacy helps individuals and organizations think across nested time horizons, from days to decades to generations. Aligning financial, governance, and behavioral signals through coherent dashboards and indices magnifies collective action. Building robust digital and physical platforms facilitates collective transitions, while fostering participatory foresight and ethical dialogue ensures that sensemaking is not centralized but distributed.

The Role of Quantum Institutions

Institutions in this quantum framework are no longer rigid structures of command and control but dynamic choreographers of change. They facilitate emergence rather than dictate outcomes, prioritize transparency over opacity, and evolve their purpose through iterative design. Humility replaces the illusion of infallibility. Such institutions must be capable of holding paradoxes, embracing plural perspectives, and co-creating coherence with the communities they serve.

The Role of Hope in a Quantum Frame

Hope, often dismissed as a naive sentiment, becomes a vital force in a nonlinear world. It is not merely an emotion but an activating principle. Hope fuels agency, sustains momentum during periods of uncertainty, and creates attractors that pull behavior toward desired outcomes. Quantum hope does not promise that the future will be good; rather, it affirms that the future remains open and that we have the power to influence it.

Becoming a Coherent Field

The ultimate invitation of this book is to move beyond adopting a theory of change and instead to become a coherent field—a living, resonant node whose alignment influences the larger system. This means practicing our principles persistently, even if imperfectly; crossing boundaries—disciplinary, national, generational—to shape institutions with care and courage; honoring the wisdom of both the Earth and our ancestors; and building communities of practice that are rich in feedback, love, and ethical awareness. In a field of incoherence, coherence itself is a form of revolution.

A Closing Vision

Imagine a world where carbon is no longer viewed as a pollutant but as a critical signal for planetary balance, where markets are tuned not to blind growth but to planetary stability, and where governance is a shared choreography rather than a top-down directive. Picture citizens not as passive consumers but as active designers of futures, where climate action is driven by meaning rather than fear, and where the future is both walked back and lived forward by millions of quantum walkers. This is not utopia—it is coherence, the true essence of a quantum walk through sustainability.

Mini Chapter

Thriving in a Nonlinear World

Strategies for Effective Sustainable Development Policy in the Face of Uncertainty

Here we discover the latest strategies and approaches for effective sustainable development policy in a world characterized by nonlinear change and uncertainty. First, let us understand the **Nonlinear Sustainability Dynamics** in further detail.

Nonlinear Sustainability Dynamics – Fundamentals

Nonlinear sustainability dynamics (**Appendix I-III**) describes how environmental, social, and economic systems change in ways that aren't simple straight lines. Instead of a steady, predictable progression, these systems can experience sudden shifts, rapid accelerations, or unexpected collapses. This is a crucial departure from many traditional approaches to sustainability, which assume linear cause-and-effect and miss the potential for abrupt transformations.

Much of this behavior is driven by feedback loops—when a system's output becomes an input that influences future states. For example, rising temperatures melt polar ice, reducing the Earth's reflectivity, which in turn accelerates warming—a classic positive feedback loop. In **Appendix XIX - Nonlinear Casuality** similar patterns are mapped across climate, energy, and resource systems. Please see for more insight.

Small changes can sometimes trigger disproportionately large effects—what frames as "low input → high output" interventions. Recognizing these tipping dynamics is key to avoiding collapse and catalyzing rapid positive shifts.

Why Linear Thinking Fails

Linear projections assume that steady effort yields steady progress. But in nonlinear systems, thresholds, feedback dominance, and time lags break this logic. For example ecological restoration may seem slow until a key species recovers, triggering a rapid ecosystem rebound. Conversely, fisheries or groundwater systems can appear stable until they suddenly crash when biological or hydrological thresholds are crossed.

Linear thinking in policy can lead to overconfidence (assuming gradual decline) or missed opportunities (failing to invest early in interventions that could trigger positive tipping points). As in **Chapter 12** (**Quantum Governance and Polycentricity**), governance must adapt to manage uncertainty and rapid shifts rather than rely on outdated predict-and-control models.

Key Drivers of Nonlinearity

- **Thresholds & Tipping Points** – Crossing a temperature or pollution threshold can set off self-reinforcing loops (e.g., permafrost methane release).
- **Positive & Negative Feedbacks** – Amplifiers like ice-albedo effect vs. dampeners like vegetation regrowth.
- **Time Lags** – Decades between emissions and climate impact; years between groundwater extraction and land subsidence.
- **Network Effects** – Localized failure in a supply chain triggering global disruption (**see Appendix IV – Real World examples – Nonlinear vs. Linear Behavior of Supply Chain…**).

System Mapping & Early Warnings

Methods like causal loop diagrams, agent-based modeling, and network analysis help identify leverage points (**Appendix XIV**) and monitor for critical slowing down—a sign of approaching tipping points **Appendix VII**, **XVIII and IX** further discuss - mitigation strategies, early warning detection enables "pre-tipping" interventions, such as expanding renewable storage before fossil phase-outs trigger instability.

Cross-Sectoral Examples

- **Cape Town's Day Zero** – A nonlinear shift from water sufficiency to crisis due to combined drought, demand growth, and infrastructure lag.
- **Amazon Dieback** – Loss of forest cover reduces rainfall, further degrading the ecosystem and impacting climate stability globally.
- **Renewable Energy Tipping** – Rapid cost declines, driven by policy and scaling feedbacks, flipping markets away from coal within a decade

Actionable Insights

- **Trigger reinforcing loops early** – e.g., feed-in tariffs or public procurement for cleantech.
- **Pair scaling with stability measures** – grid reform, storage buildout, sector coupling.
- **Integrate sectors** – so gains cascade rather than remain siloed.
- **Embed precautionary margins** – in climate targets to account for nonlinear risks like AMOC collapse.

Thriving Amid Uncertainty: Building Adaptive and Resilient Systems

In an era defined by volatility and interconnected shocks, **resilience** is not merely a policy buzzword—it is the organizing principle for sustainable progress. Nonlinear systems such as economies, ecosystems, and social networks do not evolve in straight lines. They absorb disturbances, reorganize, and adapt, often producing outcomes far from what linear models predict.

To thrive within such complexity, societies must design **policies that learn, diversify, and self-correct**—not those that attempt to predict or control every variable.

The Logic of Resilience

Resilience in sustainable development refers to a system's ability to **maintain function and purpose despite turbulence**. This requires flexible architectures—whether in energy, agriculture, or finance—that can bend without breaking.

An adaptive policy framework emphasizes:

- **Diversity** – Encouraging multiple solutions rather than dependence on one dominant path (e.g., promoting solar, wind, biomass, and storage rather than a single "silver bullet").
- **Redundancy** – Designing backup pathways for essential services so that one failure does not cascade into systemic collapse.
- **Connectivity with buffers** – Linking systems intelligently but avoiding over-integration that spreads risks too rapidly.

Such design thinking borrows more from ecology than from traditional economics: forests, for instance, survive fires not because they prevent them, but because they **evolve to regenerate afterward**.

Feedback Loops, Cascades, and Tipping Dynamics

Every sustainable system contains **feedbacks**—recursive loops where actions generate consequences that influence future actions. These loops can either stabilize (negative feedbacks) or amplify (positive feedbacks) change.

When multiple feedbacks align, **tipping points** can emerge—moments where small disturbances trigger disproportionately large transformations. Climate systems, social behavior, and innovation networks all demonstrate such nonlinear thresholds.

Policy insight: Managing feedbacks is often more effective than managing individual events. For example, subsidies for clean energy create positive feedbacks that accelerate cost reduction and adoption. Overexploitation of groundwater or deforestation can cross tipping points that abruptly degrade entire ecosystems. Thus, policymaking in nonlinear domains is less about *control* and more about *orchestration*—guiding loops to reinforce resilience and dampen collapse.

From Experimentation to Evolutionary Policy

Linear governance assumes predictability; **quantum governance** embraces iteration and learning. Effective sustainability strategies behave like **evolving organisms**: they test, adapt, and improve through feedback.

Embedding learning into policy involves:

- **Pilot experimentation** – Testing interventions on a small scale before systemic rollout.
- **Adaptive monitoring** – Using indicators and data feedback to adjust programs dynamically.
- **Institutional humility** – Recognizing that uncertainty is permanent, not temporary.

Cities such as **Copenhagen** and **Singapore** illustrate this philosophy through iterative climate and urban design programs—each cycle of feedback informing the next phase of adaptation.

Case Snapshots: Adaptive Systems in Action

Rotterdam's "Room for the River" initiative integrates green infrastructure, water storage, and public engagement—transforming flood management from a defensive stance to a regenerative urban strategy. **Kenya's mobile microfinance networks** demonstrate adaptive experimentation—leveraging feedback from rural users to refine inclusion and resilience models in real time. **Chile's flexible renewable-energy auctions** build diversity into supply portfolios, allowing the market to pivot dynamically as technologies evolve. Each of these cases illustrates resilience as *an active design process*, not a static policy outcome.

Navigating the Nonlinear Future

To navigate uncertainty, sustainable policy must internalize four guiding principles:

1. **Anticipate shocks** but avoid paralysis by prediction.
2. **Design for adaptability**—embed redundancy and modularity.

3. **Amplify positive feedbacks** that accelerate learning and regeneration.
4. **Institutionalize experimentation**—turn uncertainty into a resource, not a risk.

The goal is not merely to *survive disruption*, but to **use disruption as raw material for innovation**—transforming vulnerability into vitality.

Key Takeaway

A nonlinear world rewards **learning systems** over rigid ones. Resilience is achieved when institutions, technologies, and communities can sense, respond, and evolve faster than the shocks that threaten them. In this sense, thriving sustainably is less about resisting change—and more about **dancing with it**.

High Level Overview

Nonlinear solutions to the nonlinear climate and sustainability crisis recognize that environmental, social, and economic systems do not respond predictably or proportionally to interventions. Instead, these systems exhibit thresholds, tipping points, feedback loops, and cascading effects—meaning small actions might yield outsized effects, while conventional strategies could be rendered ineffective if system limits are crossed. It represents a departure from the Newtonian clockwise universe paradigm => the comforting illusion that cause A reliably produces effect B in a proportional manner. Instead, it embraces the unsettling reality that small inputs can trigger disproportionality large outputs (the proverbial butterfly effect), that effects can loop back to influence their causes (feed dynamics), and that system behavior can undergo abrupt, unpredictable shifts (tipping points). In the realm of sustainability, this understanding is not merely academic; it is fundamental.

Recognizing nonlinear causality compels us to adopt *systems thinking*, to anticipate feedback loops, to identify potential tipping points, and to design interventions that are adaptive and resilient in the face of unpredictability. It demands humility about our predictive capabilities and fosters a focus on building systemic robustness rather than optimizing for a single, potentially fragile, outcome.

Understanding nonlinear causality forces a critical re-evaluation of how we perceive intervention and consequences within complex environmental, sustainability and social systems.

NEWTONIAN MINDEST ⇔ EMERGENT MINDSET

Key Concepts in Nonlinear Responses

- **Tipping Points** – Critical thresholds where a small perturbation can trigger large, potentially irreversible changes (e.g., Arctic sea ice melt, coral reef die-off).

- **Feedback Loops** – Self-reinforcing processes (positive feedbacks exacerbate problems; negative feedbacks stabilize systems), such as ice-albedo feedback or methane release from thawing permafrost.
- **Cascading Effects** – Chain reactions set off by initial disruptions, leading to broad-scale impacts, like ecosystem collapses or food system failures.
- **Emergent Properties** – New behaviors or patterns arising from complex interactions that can't be predicted by analyzing parts in isolation.

The Quantum Model of Change rests on *eight interwoven principles*. Superposition allows us to hold multiple futures, values, and truths at the same time, resisting binary thinking. Entanglement reminds us that actions in one part of a system ripple through the whole, making local choices inseparable from global consequences. The Uncertainty Principle teaches us not to wait for perfect information but to make ethically grounded decisions amid ambiguity. The Observer Effect emphasizes that the act of measuring or narrating change actively reshapes outcomes, making metrics and stories equally powerful. Decoherence warns that disconnection and lack of feedback often lead to systemic collapse, while Quantum Tunneling shows that seemingly impossible transitions become feasible through focused energy and resonance. Coherence—alignment of values, actions, and signals across scales—is the organizing force, while Time Nonlinearity reveals that small present actions can influence large, delayed outcomes and, conversely, that tipping points can arrive suddenly and irreversibly.

The journey towards embracing nonlinear causality has been one of converging insights from diverse fields => chaos theory, complexity science, ecology, systems dynamics, even social sciences exploring divergent group behaviors. It reflects a growing awareness that the intricate web of life, including human interactions within it, operates according to principles far removed from simple, additive logic. This recognition carries profound implications for governance, risk management, and the very definition of sustainable development.

Quantum Map of Change

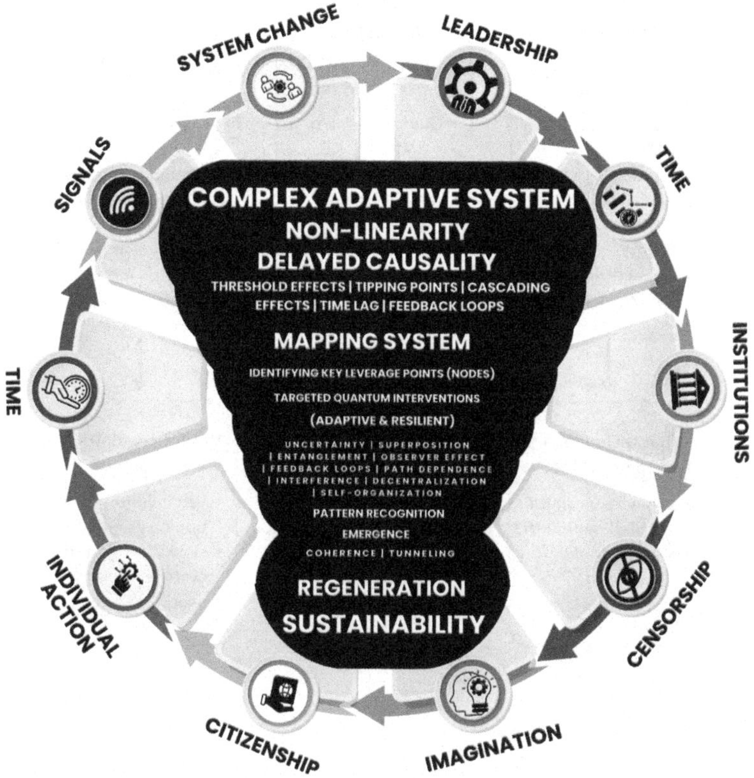

It shifts the focus from seeking singular 'silver bullet' solutions to fostering adaptive capacity and navigating emergent properties. If our interventions in complex systems inevitably ripple outwards in unpredictable ways, how then do we ethically and effectively steer towards desired long-term outcomes like planetary health and equitable human well-being?

In a quantum universe, the future is not a destination: It is a field. We walk toward it by walking from it.

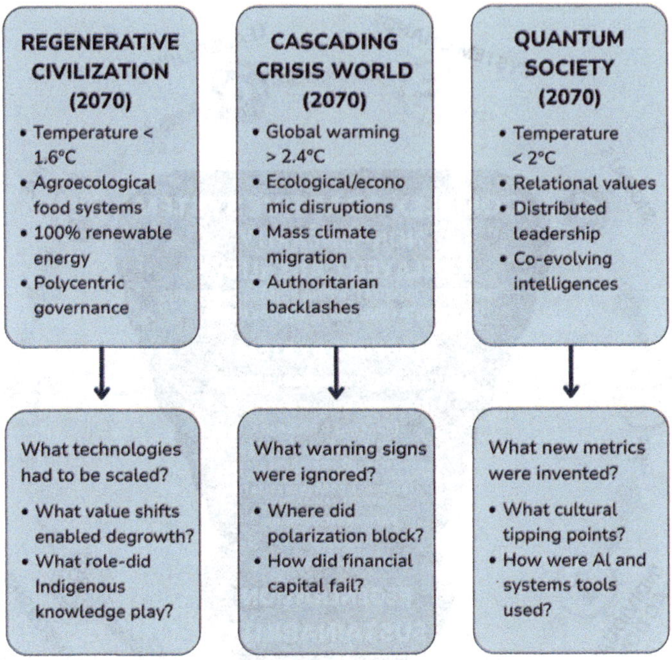
REGENERATIVE CIVILIZATION (2070)
• Temperature < 1.6°C
• Agroecological food systems
• 100% renewable energy
• Polycentric governance
CASCADING CRISIS WORLD (2070)
• Global warming > 2.4°C
• Ecological/economic disruptions
• Mass climate migration
• Authoritarian backlashes
QUANTUM SOCIETY (2070)
• Temperature < 2°C
• Relational values
• Distributed leadership
• Co-evolving intelligences
What technologies had to be scaled?
• What value shifts enabled degrowth?
• What role-did Indigenous knowledge play?
What warning signs were ignored?
• Where did polarization block?
• How did financial capital fail?
What new metrics were invented?
• What cultural tipping points?
• How were AI and systems tools used?

Epilogue

The Pulse Beneath Our Feet

There are moments in planetary time when change accelerates—not gradually, but through rupture. A threshold is crossed, and suddenly, the rules no longer hold.

What was once inconceivable becomes inevitable. What was once invisible becomes undeniable.

We are living in such a moment.

The atmosphere is no longer a background. It is an archive of our choices.

The soil is no longer inert. It is a witness to both extraction and healing.

The ocean is no longer vast and mute. It is acidifying, rising, and responding.

And the future is no longer distant. It is folding into the now—through heatwaves, migration, cloudbursts, and floods.

And yet—beneath the noise of crisis, something quieter is pulsing.

It is not panic.

It is **potential**.

Across the world, people are innovating, imagining, resisting, and reweaving.

Farmers are reviving ancestral practices. Cities are learning to breathe again. Youth are leading with moral clarity. Artists are remapping meaning. Entrepreneurs are daring to design for circularity. Elders are remembering what the land once knew.

This book was never meant to offer a singular roadmap.

Instead, it offered a **pattern language**—a quantum grammar—for reading, engaging, and shaping our moment.

It argued that sustainability is not a destination.

It is a **mode of perception**.

A way of being in time.

A commitment to coherence in the midst of chaos.

If you remember nothing else from these pages, remember this:

You are already entangled.

The only question is: *What field of futures will your coherence help bring into being?*

May you walk the future not with certainty, but with courage.

May you act not because you are sure of success, but because you are aligned with life.

And may you find, in this quantum walk, others who walk with you—not behind, not ahead, but beside.

Together, may we tune the planetary field toward regeneration, justice, and joy.

Part VII: Appendices

Appendix I

Understanding Nonlinear Dynamics in Earth's Climate System

Understanding Complexity Beyond Linearity

From Motion to Transformation

At its core, *dynamics* is the study of how things change — how states evolve when energy, force, or feedback act upon them. While it originated in classical mechanics, describing the paths of planets or the motion of springs, the same logic applies to biological evolution, market cycles, and even the Earth's climate. Every system that changes with time can be viewed as a *dynamical system*.

But unlike simple mechanical systems, natural systems rarely behave in proportion to what we do to them. Their responses to stress are not straight lines—they are *curves, spirals, thresholds, and leaps*. This is the realm of **nonlinear dynamics**. A forest may absorb carbon for decades, then collapse after one dry summer or drought; an economy may stay stable for years before a minor shock triggers global contagion. Such disproportionate responses or behaviors are not anomalies—they are the rule. Nonlinear systems are those in which **small causes may yield large effects, and history matters as much as present inputs.**

Their behavior is emergent, self-referential, and adaptive—products of interaction rather than isolation.

What Makes a System Nonlinear

In a linear world, doubling an input doubles the output — a simple, proportional cause–effect relation. But in a nonlinear world, a tiny nudge can trigger a cascade, while a massive push may do almost nothing. The relationship between cause and effect bends, folds, or sometimes fractures.

A classic metaphor is a **rubber band versus a spring.** Stretch a spring gently, and it oscillates predictably—the pull and the rebound are balanced. Stretch a rubber band past its limit, and it warps, snaps, or vibrates irregularly. Its response no longer scales linearly with force; the system has crossed into **nonlinear behavior.**

Most of nature—from weather to ecosystems, behaves like that rubber band. A forest, an ocean, or an economy rarely returns neatly to "normal" after disturbance; it reorganizes itself, sometimes into a new state entirely. The capacity for reorganization defines both their **resilience** and **fragility**.

The Climate as a Nonlinear System

Earth's climate is a vast mosaic of **nonlinearity** or interlocking nonlinear interactions. Atmospheric circulation, ocean currents, ice–albedo feedbacks, and biospheric exchanges interact in complex, coupled ways— from molecular chemistry to planetary heat flows. A subtle shift in one component—say, Arctic ice melt or rainforest evapotranspiration—can amplify or dampen others, producing **feedback loops** that can either stabilize the planet or push it toward abrupt change.

Because these processes operate across scales (from molecular to planetary) and time frames (from seconds to centuries). These operation over vastly different spatial and temporal scales, what mathematicians call **chaotic determinism**—unpredictable in detail yet patterned in form over time.

The deeper we explore the Earth's climate, the clearer it becomes that its apparent stability conceals a web of dynamic feedbacks and threshold behaviors. These interactions—between atmosphere and ocean, biosphere and cryosphere, human systems and natural processes—create patterns that defy linear prediction yet reveal underlying coherence. To understand and govern such a world, we must look beyond deterministic models and enter the realm of **nonlinear dynamics**, where resilience, feedback, and sudden change define the rhythm of both nature and society.

Examples abound across the biosphere:

- **Eutrophication in lakes** – Nutrient accumulation triggers abrupt oxygen collapse.
- **Cryosphere feedbacks** – Ice loss amplifies warming through reduced albedo.
- **Biological invasions** – Once an invader surpasses ecological resistance, it reshapes entire ecosystems.
- **Fire dynamics** – Incremental warming can convert stable forests into recurring burn cycles.

Such shifts remind us that environmental resilience has limits—and those limits often reveal themselves only after they've been crossed.

Lessons from the Lorenz Model[134]: The Butterfly Within the Climate

In 1963, meteorologist **Edward N. Lorenz** at MIT made a discovery that forever changed how we understand the atmosphere—and complexity itself. While experimenting with a simplified model of atmospheric convection using just **three equations** (variables) to represent fluid motion and temperature gradients, Lorenz noticed that even minute changes in initial conditions produced dramatically different trajectories. This became the foundation of **chaos theory**.

At a Rayleigh number near 28, his model produced looping, non-repeating trajectories in a three-dimensional "phase space." These elegant patterns, known as the **Lorenz attractor**[135], reveal how deterministic systems can behave unpredictably—bounded by structure yet infinitely varied in detail.

In this "strange attractor," the system's path never repeats exactly, but it also never escapes its domain—a paradoxical balance of **order within apparent disorder**. This finding dismantled the assumption that perfect prediction was possible if only equations were known precisely. Instead, Lorenz showed that sensitivity to initial conditions—now famously captured by the metaphor of a **butterfly's wings altering the weather** or the **"Butterfly Effect"**—makes the future inherently uncertain in complex systems like climate.

This insight not only reshaped meteorology but also became a cornerstone of modern **nonlinear dynamics**, influencing ecology, economics, and sustainability science. As explained by the Columbia University Lamont-Doherty Earth Observatory (LDEO)[136], even a three-variable model such as Lorenz's can exhibit an extraordinary range of emergent behaviors, demonstrating how small-scale perturbations can cascade into large-scale transformations in real-world systems. The Lorenz attractor remains one of the most visually striking and conceptually profound examples of structured chaos—a mirror of Earth's climate system itself.

<u>Patterns, Predictability, and the Limits of Control</u>

Nonlinear systems are not random; they exhibit **emergent order**. Weather patterns, ecological oscillations, and economic cycles often repeat in recognizable yet never identical forms. These patterns teach us that **prediction has limits, but anticipation is possible**. By identifying feedbacks and thresholds, scientists can map **zones of stability** and **zones of fragility** across the Earth system.

- **Negative feedbacks** stabilize (e.g., predator–prey cycles, radiation balance in climate).
- **Positive feedbacks** destabilize (e.g., methane release from thawing permafrost, nutrient enrichment in lakes).

When cumulative pressures push systems beyond a **tipping point**, they reorganize into a new equilibrium—a process known as a *regime shift*. Coral reefs turning to algae, wetlands drying into savannas, or ocean circulation reversing under freshwater influx exemplify this transformation. Once crossed, such thresholds are often irreversible; the system's "memory" rewrites itself.

- Ocean–atmosphere coupling can stabilize regional climates—until a tipping threshold, such as El Niño, reorganizes global patterns.
- Polar ice reflects sunlight (a cooling feedback) until meltwater absorbs heat (a warming feedback), amplifying change.

Understanding these couplings allows us to foresee possible futures, even if we cannot pinpoint exact ones.

Why This Matters for Sustainability

The same nonlinear features that make climate forecasting difficult also underpin social, economic, and ecological transitions. Sustainable development, too, is a nonlinear process. A policy reform or technological innovation may create **positive feedback loops**, accelerating adoption and innovation far faster than expected. Conversely, resource depletion or biodiversity loss may cross **critical thresholds**, leading to abrupt systemic collapse.

Recognizing nonlinearity helps policymakers and scientists move beyond simplistic "input-output" models toward **adaptive management**—where flexibility, monitoring, and feedback become the core design principles.

A Quantum View of Complexity

From a "Quantum Walk" perspective, nonlinearity is not a nuisance—it is the *engine of creativity in nature*. The universe evolves not through perfect order but through **probabilistic leaps, branching possibilities, and adaptive reconfigurations**.

In this light, sustainability itself becomes a *dance with uncertainty*: designing systems that can thrive across multiple possible trajectories, rather than betting on one stable state. It means cultivating resilience not as rigidity, but as **fluid coherence within chaos**.

Implications for Policy and Governance

Governance frameworks designed for linearity often collapses under nonlinear stress. Predictive control must give way to **adaptive learning, anticipatory design, and resilience planning.**

Key strategies include:

- early-warning indicators for approaching thresholds;
- redundancy and diversification to prevent cascading failure;
- continuous feedback integration into policy cycles;
- scenario modeling to prepare for multiple potential futures.

This shift—from rigid control to flexible adaptation—marks the next frontier of sustainability governance.

From Stability to Adaptive Flow

Recognizing nonlinearity means redefining what "stability" means. It is not the absence of change, but **the capacity to absorb disturbance and reorganize** without losing function. Sustainability, in this sense, mirrors evolution itself: a continuous process of learning, feedback, and transformation. True resilience isa achieved not through resistance, but through dynamic balance: flexible without fragmentation.

Key Takeaway

The Earth's climate, ecosystems, and human institutions all function as nonlinear, feedback-driven networks. Predictability has limits, but pattern recognition offers power. The true challenge of sustainability is not to eliminate chaos, but to *learn from it*—to build systems, economies, and societies that can evolve gracefully within the planet's dynamic flow—that can **learn faster than they break.**

Appendix II

The Role of Nonlinear Dynamics in Shaping Environmental Outcomes

Why Nonlinearity Governs Outcomes (Not Just Inputs)

Environmental systems rarely respond in straight lines. Interacting processes—biophysical, climatic, and social—create **feedbacks, thresholds, and path-dependence** that bend cause–effect. Small pressures can accumulate quietly and then, beyond a **critical threshold**, trigger disproportionate change (a regime shift). Equally, large interventions sometimes do little if countervailing feedbacks absorb them. Managing sustainability therefore means managing **system dynamics**, not just "levers" and "knobs."

<u>Core mechanisms</u>

- **Positive feedbacks** – Amplify change (e.g., ice–albedo, methane from permafrost).
- **Negative feedbacks** – Stabilize (e.g., predator–prey regulation, radiative balance).
- **Hysteresis** – Makes recovery paths different from collapse paths.
- **Multi-scale coupling (space/time)** – Links local shocks to global consequences.

Canonical Thresholds Across Earth Systems

Below are concise, policy-relevant exemplars of nonlinear behavior. They are framed to emphasize **mechanism → threshold → outcome**.

<u>Lake eutrophication (nutrient threshold)</u>

Low phosphorus keeps algae in check; gradual enrichment crosses a **trophic threshold**, oxygen collapses, fish die-offs ensue. Reversal requires more effort than prevention due to hysteresis (legacy nutrients, internal loading).

<u>**Atlantic overturning circulation (AMOC) and freshwater input**</u>

- **Mechanism** – Density-driven conveyor steered by temperature/salinity.
- **Threshold** – Sufficient freshening (e.g., Greenland melt) weakens sinking.
- **Outcome** – Abrupt regional climate re-patterning (e.g., Europe winters, monsoon shifts).

<u>**Amazon forest–rainfall coupling**</u>

- **Mechanism** – The forest recycles rain via evapotranspiration.
- **Threshold** – Deforestation near a critical share of cover undermines moisture recycling.
- **Outcome** – Savannization, biodiversity loss, and large carbon pulse.

<u>**Permafrost thaw–methane feedback**</u>

- **Mechanism** – Thaw unlocks microbially available carbon → CH_4/CO_2.
- **Threshold** – Once thaw depth and period pass a limit, warming self-reinforces.
- **Outcome** – Step-change in radiative forcing even with other emissions stabilized.

<u>**Coral bleaching**</u>

- **Mechanism** – Coral–algal symbiosis sensitive to heat exposure.
- **Threshold** – 1–2 °C sustained heat anomalies (and/or marine heatwaves).
- **Outcome** – Mass bleaching → regime shift to algae-dominated reefs; slow/uncertain recovery.

<u>**Arctic sea-ice albedo**</u>

- **Mechanism** – Reflective ice vs. absorptive open water.

- **Threshold** – Marginal warming flips local ice cover → runaway melt segment.
- **Outcome** – Amplified warming; seasonal ice loss accelerates.

<u>Aerosol masking ("cleaner air, warmer climate" paradox)</u>

- **Mechanism** – Sulfate aerosols cool by scattering sunlight.
- **Threshold** – Rapid aerosol declines without commensurate GHG cuts unmask warming.
- **Outcome** – Near-term **warming jump** despite health gains from cleaner air—an archetypal cross-sector nonlinearity.

Socio-Ecological Nonlinearities: When Humans Complete the Loop

<u>Fishery collapse archetype</u>

- **Initial** – Harvest < replenishment → stable stocks.
- **Pressure rise** – Effort increases; stocks drift down slowly.
- **Threshold** – Spawning biomass falls below recruitment needs → positive feedback (fewer spawners → fewer juveniles).
- **Outcome** – Rapid collapse; recovery requires stronger controls and long lags (hysteresis).

<u>Other sustainability examples</u>

- **Groundwater** – Slow drawdown crosses subsidence threshold → aquifer storage and quality degrade irreversibly.
- **Urban heat islands** – Incremental build-out and vegetation loss tip neighborhoods into persistent heat stress regimes.

Diagnosing Nonlinear Risk (Before It Bites)

Move from "forecasting the future" to detecting approach to thresholds.

<u>Early-warning indicators (EWIs)</u>

- **Critical slowing down** – Rising autocorrelation/variance in time series.

- **Flickering** – Intermittent jumps between alternative states.
- **Skewness changes** – Distribution tilts as resilience erodes.
- **Network fragility metrics** – Reduced modularity, loss of weak-tie buffers.

<u>Where to deploy EWIs</u>

- **Lakes and coastal zones** – Oxygen, chlorophyll, turbidity.
- **Permafrost regions** – Active-layer depth, soil moisture, ebullition CH_4.
- **Forest–savanna ecotones** – Greenness persistence, vapor-pressure deficit.
- **Power and food systems** – Reserve margins, price variance, supply-chain centrality.

Managing Nonlinear Dynamics: From Control to Adaptation

Because detailed prediction is limited, the goal is **adaptive capacity**—systems that sense, pivot, and recover.

<u>Design principles</u>

- **Diversity & redundancy** – Multiple ways to perform critical functions; modular grids; portfolio energy mixes.
- **Buffers & safe-to-fail** – Floodplains, wetlands, living shorelines; compartmentalization to prevent cascades.
- **Feedback-aware policy** – Triggers that tighten/relax measures based on monitored thresholds (index-linked policy).
- **Iterative pilots** – Small experiments → scale what works; retire what doesn't.

<u>Decision frameworks</u>

- **Adaptive Management** (monitor–evaluate–adjust cycles).
- **Robust Decision Making (RDM) and Dynamic Adaptive Policy Pathways (DAPP)** for many-futures planning.
- **No-regrets and low-regrets** actions that pay off across scenarios (efficiency, demand response, nature-based solutions).

Modeling & Measurement: Tools for a Nonlinear World

- **Agent-based & system-dynamics models** – Capture feedbacks and heterogeneity.
- **Network analysis** – To find hubs, bridges, and cascade pathways.
- **Machine learning** – Pattern detection in high-dimensional data; surrogate models for fast scenario sweeps.
- **Data assimilation** – Blending observations with models for real-time state estimation.
- **Resilience metrics** – Recovery rate, variance, redundancy indices, connectivity under stress.

Policy Implications (Mitigation & Adaptation)

- **Mitigation sequencing** – Cut GHGs and manage aerosol phase-down pacing to avoid warming jumps.
- **Carbon-ecosystem coupling** – Protect moisture-recycling forests and peatlands; avoid crossing biophysical thresholds that unlock carbon pulses.
- **Infrastructure timing** – Bring storage, demand flexibility, and interconnection online before retiring dispatchable capacity—avoid reliability cliffs.
- **Equity lens** – Nonlinear shocks hit vulnerable communities hardest; build social buffers (insurance access, heat refuges, early-warning reach).

Planning for Development Goals Under Nonlinearity

Sustainable development requires systems literacy:

- map interaction webs (economy–energy–water–land–health);
- identify leverage points (pricing, standards, social norms) that drive positive feedbacks (e.g., EV adoption curves);
- use portfolio approaches (many small bets > one big bet);
- bake **contingencies** into program finance (adaptive tranches tied to EWIs).

From Pressure to Pivot

A single schematic showing: **pressures accumulate → resilience erodes (EWIs rise) → threshold crossed → new regime**; alongside a policy pathway with triggers, buffers, and options

Challenges You Should Expect (and How to Turn Them)

- **Challenges** – uncertainty, hidden thresholds, complex attribution, policy lag.
- **Opportunities** – earlier detection, targeted buffers, faster learning cycles, and **positive tipping** (e.g., clean-tech S-curves via incentives + social diffusion).

Practical Checklist (One Page, Action-Oriented)

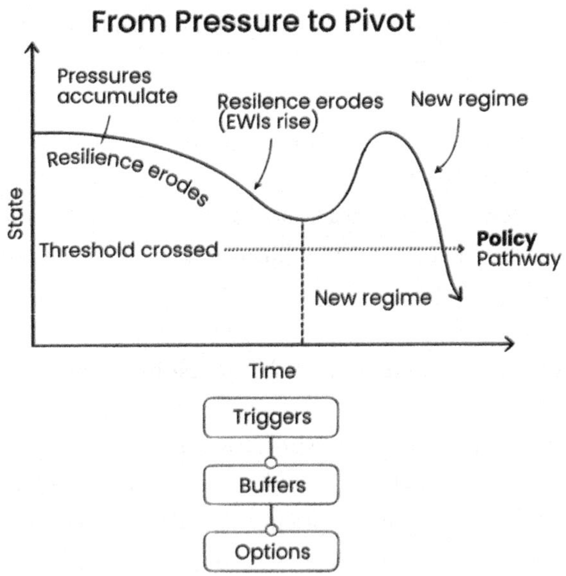

Figure II.1

Diagnose

- track EWIs (variance, autocorrelation, flicker);
- stress-test with multi-futures scenarios (RDM/DAPP);
- map feedbacks & tipping elements (biophysical + socio-technical).

Design

- add buffers (storage, green infrastructure, insurance);
- build redundancy & modularity (microgrids, diversified supply);
- set automatic policy triggers (index-linked adjustments).

Deliver

- pilot → evaluate → scale (time-boxed cycles);
- publish dashboards; socialize signals with communities;
- pre-commit resources for rapid pivot when signals trip.

Key Takeaway

Environmental and sustainability outcomes are properties of system dynamics, not just policy inputs. Progress accelerates when we amplify constructive feedbacks, detect weakening resilience early, and design institutions to adapt on the fly. In a nonlinear world, success belongs to systems that learn faster than they fail.

Appendix III

How Do Nonlinear Dynamics Influence Ecosystem Resilience and Management Approaches?

Nonlinear dynamics fundamentally shape both the resilience of ecosystems and the strategies used to manage them. Here's how:

- **Thresholds and Tipping Points** – Nonlinear ecosystems can experience abrupt transitions—small changes (like a bit more pollution or reduced rainfall) can drive a sudden shift to a new regime, often with reduced ecosystem function or biodiversity. This means systems may appear stable until a critical threshold is crossed, after which recovery to the previous state can be extremely difficult or impossible.
- **Alternative Stable States** – Ecosystems may have multiple stable conditions (e.g., clear vs. turbid lakes). Crossing a threshold can lock the system into a less desirable state, maintained by its own self-reinforcing feedbacks, rather than returning to its original condition even if stressors are reduced.
- **Feedback Loops** – Both positive (amplifying) and negative (stabilizing) feedbacks drive nonlinear behaviors. For example, in coral reefs, loss of cover can rapidly shift balance toward algae, fundamentally altering ecosystem dynamics. Such feedbacks are central to resilience, as they determine how systems respond to disturbances.
- **Emergence and Complexity** – Ecosystem resilience is an emergent property arising from complex interactions among species, physical processes, and their environment. These interactions can sometimes buffer disturbances but can also amplify them unexpectedly due to the networked nature of ecological systems.

Implications for Ecosystem Management Approaches

- **Need for Adaptive Management** – Because of unpredictability, management needs to be flexible and experimental—constantly monitoring, learning, and adjusting strategies as ecosystems respond in unexpected ways. Rigid, control-oriented approaches often fail in the face of nonlinear change.
- **Emphasize Biodiversity and Connectivity** – High diversity and strong connectivity between habitats promote redundancy and adaptability, allowing ecosystems to better absorb shocks and reorganize after disturbance. These are priority targets for resilience-focused management.
- **Precautionary Principle** – Since the consequences of crossing ecological thresholds can be severe and irreversible, ecosystem managers employ precaution—erring on the side of caution, especially near potential tipping points.
- **System Health, Not Single Outputs** – Management goals shift from maximizing isolated outputs (like timber or fish) to maintaining overall system health, structure, and function over the long term. This holistic approach recognizes that the loss of resilience can threaten all services and benefits provided by ecosystems.
- **Recognition of Path Dependence (Hysteresis)** – An ecosystem's history matters. Once a threshold or regime shift occurs, simply removing the initial driver may not restore prior conditions—requiring different, sometimes costly interventions.

Examples

- **Coral Reefs** – Rising temperatures and acidification can push reefs to a tipping point, causing rapid collapse to an algae-dominated state that is difficult to reverse, regardless of subsequent improvements in temperature or chemistry.
- **Lakes** – Shallow lakes exposed to nutrient pollution may rapidly shift from clear to turbid states (dominated by algal blooms)

and remain locked in that state due to reinforcing feedbacks like decreased plant cover.

- **Forests and Grasslands** – Changes in rainfall and grazing can lead to nonlinear shifts in vegetation types—sometimes resulting in irreversible desertification depending on system history and feedbacks.

Summary Table III.1 – Key Nonlinear Features and Management Shifts

Nonlinear Feature	**Management Response**
Thresholds/Tipping Points	Precaution, monitoring, early intervention
Feedback Loops	Foster negative (stabilizing) feedbacks
Alternative Stable States	Accept possible irreversibility, target resilience
Emergence/Complexity	Holistic, system-wide perspectives
Uncertainty/Path Dependence	Adaptive management, maintain diversity

In essence, nonlinear dynamics make ecosystem resilience a moving target, susceptible to abrupt and hard-to-predict shifts. Effective management in this context is adaptive, holistic, risk-aware, and deeply mindful of both the complexity and limits inherent in ecological systems.

Appendix IV

Real World Examples

Nonlinear (Norm) and Linear (Exception) Behavior of Supply Chain, Consumer Behavior, Innovation Networks, and Climate Emissions (not Behave Linearly)

Nonlinear Behavior (the norm across all four domains)

<u>Supply Chains</u>

Supply chains are inherently nonlinear because one small disruption can cause cascading failures. Inventory, demand, and logistics are interdependent, with lag times and amplified ripple effects, while relationships are distributed and adaptive rather than centralized.

Example – COVID-19 and the Semiconductor Shortage: A single factory shutdown in Taiwan in 2020 triggered global ripple effects months later, delaying auto production, creating smartphone shortages, and raising prices in consumer electronics. Car manufacturers that relied on "just-in-time" inventory systems could not recover quickly, losing billions in revenue. The *bullwhip effect*—where small changes in demand amplify upstream—illustrates how nonlinear propagation works.

Key takeaway: Linear forecasting failed because it ignored delayed feedback, interdependence, and dynamic response.

<u>Consumer Behavior</u>

Consumer choices emerge through social norms, identity, peer effects, advertising, and context—not simply price or utility. Small nudges can lead to disproportionate changes, or none at all, and behaviors evolve path-dependently through feedback and imitation.

Example – Plastic Bag Bans vs. Default Nudges: In many regions, a 5-cent charge on plastic bags cut usage by 70–90%. The tiny cost mattered less than the signal: a shift in norms that made plastic use socially disapproved. Conversely, large subsidies for electric vehicles have often failed when supportive infrastructure or social reinforcement was absent.

Key takeaway: Behavior is not linear; it is shaped by identity, friction, and feedbacks rather than cost alone.

Innovation Networks

Innovation unfolds as a nonlinear, networked ecosystem rather than a simple pipeline from R&D to market. It is iterative, shaped by spillovers, recombination, serendipity, and cultural narratives, with breakthroughs often coming from unexpected domains.

Example – mRNA Vaccines: The COVID-19 vaccines (Pfizer-BioNTech, Moderna) emerged from decades of research in unrelated fields—lipid nanoparticles (from cancer), immune suppression (from HIV), and AI-enabled protein folding. Traditional linear planning models could not have predicted these recombinations. Similarly, network effects and first-mover narratives (Tesla, OpenAI) often determine success, not technical superiority.

Key takeaway: Innovation behaves more like a mycelial web—emergent, recursive, and nonlinear—than a pipeline.

Climate Emissions

Emissions are embedded in cultural patterns, infrastructure, pricing systems, and politics, not just fossil fuel combustion. Feedback loops like permafrost thaw or afforestation dynamically change carbon sinks and sources, while policy impacts are delayed.

Example – The Rebound Effect: Fuel-efficient cars reduce cost per mile, but this often encourages more driving, offsetting expected savings. The rebound illustrates how linear efficiency assumptions miss systemic adaptations.

Key takeaway: Emissions result from interacting systems—urban design, finance, habits, agriculture—not simply energy technology.

Linear Behavior (exceptions where causality applied)

<u>**Supply Chains**</u>

Example – Toyota's Just-in-Time (JIT) System: In Toyota's lean domestic production system, customer orders linked directly to inventory. Stable logistics and low external shocks allowed inputs, processes, and outputs to flow predictably.

When it works: Tight control, local stability, short time horizons.

Verdict: Nonlinearity dominates in global, multi-node, shock-prone supply chains.

<u>**Consumer Behavior**</u>

Example – Tobacco Taxes: In Australia and the UK, consistent tax increases reduced smoking at predictable rates. Price elasticity of demand worked reliably, with 10% price hikes leading to 4–6% declines. Public health campaigns reinforced this linear causal chain.

When it works: Habitual, price-sensitive behaviors with strong incentives and little social complexity.

Verdict: Mixed—linear for some habits, but nonlinear dominates where identity and norms matter (e.g., veganism, EV adoption).

<u>**Innovation Networks**</u>

Example – NASA's Apollo Program: The moon landing followed a linear roadmap: **basic research → R&D → prototype → test → deployment**. Hierarchical funding and a singular mission goal enabled predictable progress and spinoffs like satellite navigation.

When it works: Mission-driven, centralized, well-funded programs.

Verdict: Linear models succeed in "Big Science" or defense, but most innovation is nonlinear.

<u>**Climate Emissions and Interventions**</u>

Example – The Montreal Protocol: Banning CFCs led directly to reduced emissions and ozone recovery. Success was possible because one chemical class was responsible, alternatives existed, and global cooperation was swift.

When it works: Singular, controllable pollutants with clear substitutes.

Verdict: Nonlinear overwhelmingly dominates in climate mitigation, where multiple sources and delays prevail.

These are summarized in **Table IV.1** below.

Summary Table IV.1 – Examples of Linear Behavior

Domain	Example Where Linear Worked	When It Works	Dominant in Real Life
Supply Chains	Toyota's JIT system	Stable, local, well-controlled	❗ Nonlinear
Consumer Behavior	Tobacco tax impact on smoking	Habitual, price-elastic behaviors	⚖️ Mixed, Nonlinear tends to dominate
Innovation Networks	NASA's Apollo Program	Mission-driven, centralized R&D	❗ Nonlinear
Emissions/ Climate	Montreal Protocol and ozone recovery	Singular pollutants, clear substitutes	❗ Nonlinear

Closing Thought

We do not live in a Newtonian, clockwork world of predictable causality. Instead, we inhabit a quantum-like reality of entangled effects, hidden thresholds, and delayed responses. Our cognitive tools—spreadsheets, flowcharts, equilibrium models—often misrepresent this complexity.

Final Insight: Linear causality is the exception, not the rule. It works in structured, bounded, low-variance systems. But sustainability and climate challenges unfold in messy, adaptive, time-sensitive realities. Here, nonlinear thinking is not optional—it is essential.

Appendix V

Distributed Causality

Why No Single Actor Is to Blame

In complex adaptive systems, causation is not linear or singular. Outcomes emerge from multiple interactions, often spread across time, space, and domains. This is what we mean by "distributed causality." It is one of the most counterintuitive but crucial insights when working with complex adaptive systems. It means that no single actor, variable, or intervention fully explains outcomes. Instead, causation is spread across multiple nodes, interactions, and time horizons. Here's an expansion with examples across domains you often touch in the book:

Climate Systems

Heatwaves in Europe

Not caused by one factor (e.g., emissions from a single region), but by the interaction of: Arctic ice melt disrupting the jet stream, soil moisture loss in Eurasia, fossil fuel–driven global warming, and regional land-use changes. No single variable "causes" the heatwave—it emerges from distributed dynamics.

Deforestation in Amazon

Loss of rainfall patterns in São Paulo is linked not only to local emissions or logging, but to moisture recycling disrupted across the entire Amazon Basin. The causal web spans thousands of kilometers.

Climate Extremes

A European heatwave is not caused by "emissions alone" but by Arctic ice loss, disrupted jet streams, soil moisture feedbacks, and land-use changes acting together.

Urban and Social Systems

Traffic Congestion

Not caused only by "too many cars," but distributed across factors: land-use zoning, ride-hailing app incentives, public transit investments, cultural preferences, and even weather events.

Air Pollution (Smog Episodes) in Delhi

Attributed to local cars or industry, but causality also includes stubble burning in Punjab, cross-border emissions due to transport of pollutants from neighboring states, weather inversions, and governance failures.

Economic & Financial Systems

2008 Financial Crisis

Not caused by one "bad product" or "subprime mortgages alone," but distributed across securitization practices, global capital flows, credit rating agency incentives, monetary policy, and consumer borrowing behavior.

Carbon Price Volatility

Doesn't result only from "supply/demand of allowances." It reflects climate policy shifts, technological breakthroughs (like cheap solar), investor sentiment, and geopolitical energy shocks.

Agriculture & Food

Food Insecurity (Famine in Horn of Africa)

Not a linear result of "drought." Causality is distributed across climate variability, global food prices, local governance, international aid timing, conflict, and infrastructure fragility.

Soil Degradation

Driven by fertilizer overuse, monoculture, mechanization, global trade pressure, and climate variability—no single lever explains the collapse.

Implications | Why It Matters

- **Policy implication** – Single-point fixes rarely work (e.g., banning plastic straws doesn't address systemic waste).
- **Governance implication** – Responsibility is diffuse (who "caused" a flood—the developer who built on wetlands, the municipality that zoned it, the national climate policy that delayed action?).
- **Ethical implication** – Distributed causality challenges accountability but also demands distributed solutions.

Distributed causality makes *accountability diffuse* but action shared. It means we must design responses that embrace entangled responsibility—across states, firms, communities, and individuals—rather than searching for a single villain or silver-bullet fix.

Appendix VI

Time Flows Strangely

Nonlinear and Non-Deterministic

In Newton's universe, time is absolute: it ticks forward evenly, like the swing of a pendulum. Causes precede effects in neat sequence, and the future can be forecast if you know the present conditions. But in quantum physics—and in complex adaptive systems—time does not behave so predictably.

Nonlinear Time

- **Quantum Example** – In quantum tunneling, a particle can seem to cross an energy barrier instantaneously, as though time "short-circuits."
- **Systemic Analogy** – In climate and social systems, change often looks slow until a threshold is crossed—then it accelerates rapidly. Example: Arctic ice melt seemed incremental for decades, but now sudden losses occur in compressed timescales.

Non-deterministic Time

- **Quantum Example** – Radioactive decay happens probabilistically. We can predict half-lives for a population of atoms, but we cannot say when a single atom will decay.
- **Systemic Analogy** – Policy shifts behave similarly. The fall of the Berlin Wall, or the sudden global adoption of smartphones, could not have been timed with certainty—only described as increasingly probable. In sustainability, a carbon tax may linger politically impossible for years, then pass overnight when social, economic, and cultural forces align.

Feedback-Entangled Temporality

- **Quantum Example** – Some interpretations of quantum mechanics suggest retrocausality—where effects can appear to influence causes.
- **Systemic Analogy** – In human systems, anticipation of the future reshapes the present. For instance, when markets expect a carbon price, firms start decarbonizing before the law is passed. Future imaginaries—like the vision of a "net zero 2050"—loop backward to influence current investments.

Time Compression and Expansion

- **Quantum Example** – Particles in superposition do not evolve in a single trajectory of time but across many possible pathways until observed.
- **Systemic Analogy** – Decarbonization timelines are compressed when multiple solutions converge (solar + storage + policy + finance), and they can stretch out when barriers stack. India's renewable rollout, for example, leapt from marginal to mainstream in under a decade, while nuclear waste management remains unresolved after 70 years.

Why This Matters

For sustainability, this means **we cannot treat time as a neutral container for action.** Some solutions demand early intervention because delay multiplies risks (e.g., permafrost thaw). Others require patience for social and ecological systems to adapt (e.g., regenerative farming). Still others arrive as "punctuated moments" when windows of opportunity suddenly open (e.g., global pandemic reshaping work patterns).

Thus, time in complex systems is not a metronome but a jazz improvisation—syncopated, surprising, and deeply contextual.

Appendix VII

What Roles Do Tipping Points Play in Nonlinear Climate and Sustainability Solutions?

Tipping points are fundamental to understanding and designing effective nonlinear climate and sustainability solutions because they represent critical thresholds where small, incremental changes can cause abrupt, large-scale, and often irreversible shifts in environmental systems. Their central roles include:

Defining System Vulnerability

Tipping points highlight how certain systems (like ice sheets, rainforests, or coral reefs) can tolerate gradual stress to a limit, after which they rapidly tip into a radically different—and often less desirable—state. For example, the collapse of the West Antarctic ice sheet could raise global sea levels by several meters, and die-off in the Amazon rainforest could drastically accelerate global warming due to lost carbon storage.

Driving Nonlinear Change

The climate system is governed by interconnected feedback loops—some positive (amplifying change), some negative (dampening it). Tipping points often coincide with self-reinforcing feedbacks, such as Arctic ice melt reducing reflectivity (albedo) and causing further warming, or thawing permafrost releasing methane and boosting greenhouse gas concentrations. These nonlinearities mean risks can accelerate unexpectedly and solutions must be equally dynamic.

Guiding Mitigation Strategies

Tipping points urge early, precautionary action instead of gradual, incremental responses. Incorporating tipping points into models and policy shifts the focus to prevention (ensuring we don't cross dangerous thresholds), the development of early warning systems, and investment in resilience—building the capacity to withstand shocks and adapt fast if thresholds are neared.

Shaping Sustainability Pathways

For sustainability, tipping points shrink the room for error; once a system crosses its threshold, restoration may be impossible. This "shrinking playing field" means acting sooner and more boldly is necessary to avoid irreversible harm. This perspective encourages transformative, rather than incremental, policies—such as aggressive emission cuts, rapid ecosystem restoration, and systemic changes in land and resource use.

Unlocking Positive Transformations

Not all tipping points are negative. Positive tipping points—like the rapid uptake of renewables or widescale ecosystem restoration—can accelerate solutions through reinforcing feedbacks in technology diffusion or social norms, leading to faster systemic shifts towards sustainability.

In summary

Tipping points are both warning signals and strategic guideposts within nonlinear climate and sustainability approaches. They explain why change may be sudden, why urgent and transformative solutions are essential, and how both threats and opportunities for rapid improvement can emerge from small, well-timed interventions. Recognizing and anticipating tipping points makes climate and sustainability solutions more robust, prevention-focused, and equipped for real-world complexity.

Appendix VIII

How Might Recognizing Tipping Points Reshape Long-Term Sustainability Goals?

Recognizing tipping points can fundamentally reshape long-term sustainability goals by introducing urgency, adaptability, and transformative ambition into planning.

Shift from Incremental Improvement to Preventive Action

When tipping points are acknowledged, long-term goals prioritize *avoiding critical thresholds*, focusing on early intervention rather than gradual or delayed progress. This drives policies and investments toward preemptive solutions instead of waiting for crises to escalate.

Redefining "Safe Operating Space"

Sustainability goals become centered around identifying and respecting ecological and climate boundaries. Instead of setting targets based solely on linear projections or optimal resource use, goals emphasize staying well within known system limits to avoid abrupt, irreversible changes.

Increased Emphasis on Resilience

Recognizing that systems can rapidly shift if tipping points are crossed leads planners to build resilience at all scales. Goals incorporate diversification, redundancy, and flexibility to better adapt to shocks and reorganize after disturbances, making sustainability much more robust.

Strategic Focus on Leverage Points

Long-term ambitions target *key leverage points*—areas in social, economic, or ecological systems where modest intervention can have outsized, system-wide effects. Prioritizing such nodes increases the likelihood of preserving stability and catalyzing positive transformations.

Accelerated Timelines and Ambition

Awareness of tipping points compresses timelines for action, requiring faster progress and bolder goals. Instead of gradual transitions, sustainability objectives become time-sensitive, aiming to transform systems before thresholds are neared.

Holistic and Integrated Planning

Tipping point recognition encourages systems thinking—goals address interdependencies, feedback loops, and the potential for cascading effects. Sustainability plans expand beyond single-sector or isolated targets, embracing complexity and cross-sector integration.

Embracing Transformational Change

Long-term goals incorporate pathways for rapid, systemic shifts (e.g., rapid decarbonization, ecological restoration) rather than relying solely on incremental improvement. This leverages positive tipping points—such as the swift diffusion of sustainable technologies or practices—to accelerate progress.

Conclusion

Integrating tipping points into sustainability planning transforms long-term goals from cautious, incremental improvement to proactive, resilient, and system-wide transformation. It reframes success as not only achieving targets, but also deliberately steering away from abrupt, irreversible losses—making sustainability ambitions both safer and more achievable in a complex, dynamic world.

Appendix IX

What are Practical Methods to Detect Early-Warning Signs of Nonlinear Climate Shifts?

Practical methods for detecting early-warning signs of nonlinear climate shifts—such as abrupt changes and tipping points—rely on a mixture of statistical indicators, remote sensing, modeling, and network analysis. Here are the most established and emerging techniques:

Critical Slowing Down Indicators

- **Autocorrelation** – As a system nears a tipping point, it recovers more slowly from perturbations. This "critical slowing down" appears statistically as an increase in the lag-1 autocorrelation in time series data.
- **Variance** – Variability (variance) in climate data, such as temperature or ice cover, typically increases as a tipping point is approached.
- **Skewness and Kurtosis** – The time series distribution's skewness and kurtosis may also change, reflecting shifts in the stability or asymmetry of fluctuations.
- **Composite Early Warning Indicators (EWIs)** – These combine several statistical measures—such as autocorrelation, variance, skewness—to provide a more robust early warning signal across different regions or systems.

Time Series and Trend Analysis

- **Detrended Fluctuation Analysis** – Used to distinguish long-term correlations from random noise, helping to reveal critical slowing down and other precursor signals of tipping points.

- **Flickering** – Look for alternations between system states before a final, persistent shift—these can signal nearing a climate tipping point (e.g., alternating wet and dry periods before a climate regime shift).

Climate Network Analysis

- **Network-Based Indicators** – By creating complex networks from climate variables across spatial grids, researchers track changes in topological features (e.g., degree, betweenness centrality, average path length) that precede system-wide tipping events. This method integrates both spatial and temporal data and has proven effective in model simulations and some real-world applications.
- **Climate Subsystem Networks** – Applied to ocean and atmospheric systems (e.g., Atlantic Meridional Overturning Circulation or El Niño), these can pinpoint regions where tipping is likely to initiate.

Remote Sensing and Earth Observation

- **Satellite Monitoring** – Detects large-scale spatial changes (such as declining ice sheet integrity or ecosystem fragmentation) in near real-time.
- **Combined Observations** – Integrating satellite data with ground measurements and Earth system models to anticipate approaching critical transitions.

Machine Learning & Artificial Intelligence

- **Neural Networks** – AI methods are being trained on large climate datasets to recognize the subtle, multidimensional patterns that precede phase transitions or critical shifts.

Scenario-Based Modeling

- **Multi-Model Approaches & Ensemble Forecasting** – Stress-testing climate models under various scenarios to reveal potential points or regions where early warning indicators may spike, flagging vulnerabilities before thresholds are reached.

Example Applications

- **Greenland and Antarctic Ice Sheets** – Tracking increases in autocorrelation and variance in melt rates as indicators of increased fragility.
- **AMOC Collapse** – Monitoring natural fluctuations in the Atlantic Meridional Overturning Circulation for lengthening persistence (another form of critical slowing down), providing warnings centuries before collapse in models.

Table IX.1 – Summary Table - Practical Early Warning Methods

Method	Indicator Example	Application Area
Critical slowing down/statistics	Autocorrelation, Variance	General/regional climate data
Time series trend analysis	Detrended Fluctuation, Flickering	Paleo and observational records
Climate networks	Degree, Betweenness, Path length	Spatial grid climate datasets
Remote sensing	Ice cover, Vegetation changes	Earth observation streams
AI & Neural networks	Learned precursors	Complex climate datasets

Conclusion

Practical early warning of nonlinear shifts in the climate system depends on detecting increasing autocorrelation ("slowing down"), variance, flickering, and changes in network structure via time series analysis, network modeling, and satellite data, with AI methods showing promise for enhanced detection. Combining these tools makes it possible to spot precursors of abrupt changes, but rigorous testing and integration into monitoring systems remain active areas of research.

Appendix X

What Is a Tipping Cascade?

A tipping cascade occurs when a change in one subsystem or region triggers subsequent changes in others—**not in a simple linear fashion, but through contagion, feedback, and resonance**. Once one part of the system tips, it raises the probability or accelerates the timing of others tipping.

This concept is especially relevant to sustainability because climate, economics, social norms, and technologies are deeply interlinked. Behavioral tipping points can unlock new political or market realities, while technological adoption often accelerates once a critical mass is reached in one place.

Examples of Local Innovations Scaling via Tipping Cascades

Renewable Energy Policy in Germany (Energiewende) → Global Solar Boom

Germany's early subsidies for solar PV created manufacturing scale, driving down global prices. This price drop made solar viable in new markets, spreading adoption rapidly across China, India, Africa, and the U.S.

Cascade: Local policy → global price drop → global adoption tipping point.

Plastic Bag Bans in Cities → National Legislation

City-level bans, such as in San Francisco, provided templates and shifted consumer norms. Their success and visibility catalyzed similar policies across multiple countries, including Kenya and India.

Cascade: Local law → norm shift → political contagion → national policy.

Meatless Monday Campaigns → Mainstream Dietary Shifts

School- and city-led campaigns normalized plant-based meals. This cultural shift spurred product innovation (e.g., Impossible, Beyond), influenced restaurant menus, and prompted corporate pledges.

Cascade: Local advocacy → market innovation → industry-wide shift.

Eco-Village or Regenerative Farming Pilots → Policy and Practice Diffusion

Initiatives such as Auroville in India and permaculture farms inspired replication worldwide. NGOs and governments adopted and scaled these integrated farming models.

Cascade: Community experimentation → transnational learning → institutional uptake.

Carbon Pricing and Dividend Models

Canada's carbon tax with rebates, along with pilot efforts such as the Alaska Permanent Fund, legitimized carbon dividend proposals in the U.S. and Europe.

Cascade: Successful model → intellectual legitimacy → policy replication pressure.

Why It Matters

Recognizing the dynamics of tipping cascades helps us act more strategically. It enables us to:

- invest in leverage points that have contagion potential;
- design scalable interventions that incorporate feedback loops and visibility;
- appreciate that systemic transformation often begins with small but well-placed actions.

Appendix XI

Cascading Effects and the Management of Nonlinear Climate Risks

Understanding Cascading Dynamics

Cascading effects describe a chain of linked disruptions where a single perturbation propagates through interconnected systems—ecological, climatic, economic, or social. Unlike isolated shocks, cascades amplify because one failure becomes the input for the next. In a tightly coupled planet, local disturbances can scale into global crises.

For instance, a **heatwave** that triggers **crop failure** in one region can drive food price spikes, destabilize energy markets (via biofuel demand), and incite **migration pressures** in vulnerable zones. Similarly, **ice-sheet melt** influences ocean circulation, which alters monsoon rainfall and in turn affects agricultural productivity across continents. These interdependencies reveal that climate risk is **nonlinear, systemic, and contagious.**

Cascades are often invisible until thresholds are crossed—a reservoir drying, a coral reef collapsing, or a supply chain breaking. Once feedbacks align, recovery becomes difficult because each system reinforces the collapse of the next.

Categories of Cascading Effects

<u>Environmental Cascades</u>

- **Cryosphere feedbacks** – Arctic ice loss accelerates warming → permafrost thaw → methane release → further warming.
- **Ocean acidification** – CO_2 uptake lowers pH → shellfish decline → food web disruption → reduced oceanic carbon sink capacity.
- **Deforestation** – forest loss weakens rainfall recycling → drought → fires → more loss.

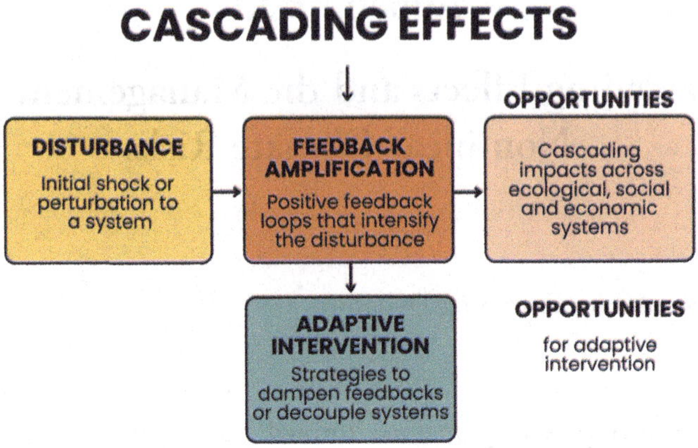

Figure XI.1: An Illustration of Cascading Effects

<u>Socio-Economic Cascades</u>

- Drought → crop failure → food scarcity → social unrest → migration.
- **Energy system coupling** – Extreme weather disables grid infrastructure → fuel supply shortages → industrial slowdown.
- **Financial contagion** – Carbon-intensive assets devalue → capital flight from vulnerable economies → inequality surge.

<u>Cross-Domain Cascades</u>

These occur when ecological stress merges with human systems. For example, flooding that damages cropland can also interrupt global supply chains and insurance systems, compounding losses. Climate shocks today are *networked phenomena*, not isolated disasters.

Why Cascades Matter for Mitigation

Linear policy frameworks—those that assume proportional cause and effect—often underestimate the **speed and scale** of climate responses. Nonlinear cascades create multiplier effects: one unmitigated tipping point can unlock several others.

Therefore, managing emissions is no longer just about reducing carbon flows; it's about **breaking feedback chains**. A mitigation policy that overlooks interconnections may inadvertently shift risk rather than reduce it (e.g., biofuel expansion causing deforestation).

Managing Nonlinear Responses: From Anticipation to Adaptation

Resilience Intelligence

An effective climate strategy integrates sensing, flexibility, and coordination. *Resilience intelligence* refers to institutional capacity to interpret early signals, test adaptive options, and respond before cascades spiral. It builds on the logic of adaptive management but expands to include predictive analytics, digital monitoring, and cross-sector communication.

Core components

- **Continuous monitoring** of early-warning indicators (soil moisture, permafrost temperature, food trade volatility).
- **Iterative learning** – Policies evolve with observed feedback, not after crises.
- **Multi-stakeholder collaboration** – Blending local ecological knowledge with scientific modeling.

Adaptive and Flexible Mitigation Frameworks

Adaptive management treats policy as an experiment: plan → implement → observe → adjust. Its goal is to maintain functionality despite uncertainty. Applied to climate mitigation:

- diversify energy pathways (renewables + storage + efficiency);
- maintain flexibility in carbon pricing, subsidies, and investment timelines;
- encourage local innovations (microgrids, urban cooling, regenerative farming) as decentralized buffers.

Modeling Nonlinear Behavior in Climate Systems

Traditional linear climate models underestimate thresholds and feedbacks. To improve realism:

- use **ensemble simulations** to bracket uncertainty ranges;
- include **bifurcation analysis** to detect potential tipping regimes;
- incorporate **data-driven learning models** (AI/ML) to capture hidden nonlinearities;
- couple **biogeophysical** and **socio-economic** layers to understand cross-sector effects.

Building Resilience Through Diversified Strategies

No single mitigation tool can address all nonlinearities. Redundancy—often viewed as inefficiency—is essential for stability.

Key practices:

- **diversify** mitigation actions across energy, agriculture, transport, and land use;
- **invest in nature-based buffers** (wetlands, forests, mangroves) to absorb shocks;
- **design flexible financial mechanisms** that can reallocate resources dynamically during crises.

Strategic Framework for Managing Nonlinear Climate Responses

Table XI.1 – Summary Table

Strategy	Core Description	Key Benefits
Resilience Intelligence	Integrates monitoring, feedback, and learning to manage uncertainty proactively	Anticipation, agility, early intervention
Adaptive Management	Iterative policy adjustment through data-driven learning cycles	Responsiveness, continuous improvement
Nonlinear Climate Modeling	Embeds thresholds, feedbacks, and multiple equilibria into prediction frameworks	Higher accuracy, realistic scenario testing
Diversified Mitigation	Portfolio of solutions across scales and sectors to distribute risk	Robustness, redundancy, cross-sector resilience

Linking Mitigation to Adaptation

Nonlinear mitigation is inseparable from adaptation. Actions that reduce vulnerability today (e.g., restoring wetlands, decentralizing energy systems) can simultaneously slow future feedbacks. *Resilience begets mitigation.*

Examples

- **Restored mangroves** – Sequester carbon *and* buffer coastal floods.
- **Regenerative agriculture** – Enhances soil carbon *and* stabilizes local hydrology.
- **Circular economy transitions** – Reduce emissions *and* shorten supply chains, lowering systemic fragility.

The Path Forward: Governing in a Nonlinear World

Climate cascades demand governance that thinks like the systems it governs: **complex, adaptive, and interconnected.**

Priorities for the next decade include:

- embedding **tipping-point literacy** in policymaking and finance;
- scaling **AI-enabled resilience** modeling for real-time scenario stress-testing;
- incentivizing **cross-sector contingency planning** (energy–water–food–health);
- fostering **distributed leadership** – empowering local actors within global frameworks.

Conclusion

Linear planning fails in a nonlinear world. Cascading effects teach us that mitigation and adaptation cannot be separated; they are co-evolutionary processes. By developing *resilience intelligence*—systems that sense, learn, and pivot—we can manage uncertainty before it manages us. The next era of climate policy will not be about prediction, but about **preparedness, feedback fluency, and adaptive design.**

Appendix XII

"Tracking 'Renewable Capacity Installed' Ignores Intermittency and Ecological Footprints"—What It Means

Governments, agencies, and energy analysts often use *installed renewable capacity* (measured in megawatts or gigawatts) as a key success metric for climate action. While this signals potential power output, it overlooks two crucial realities: **intermittency** and the **ecological footprint of renewables**.

Intermittency (or Variability)

Solar and wind are not constant energy sources—they only generate electricity when the sun shines or the wind blows. For example, a 100 MW solar farm might average only 20–25 MW of output because of night-time and cloudy days. This relationship between actual and theoretical generation is known as the *capacity* factor.

A high installed capacity number, therefore, does not guarantee consistent or dispatchable power, which is essential for grid stability. Tracking only capacity can overstate decarbonization progress unless it is supported by storage systems (such as batteries or pumped hydro), grid flexibility measures (like demand response and forecasting), or firm clean power (such as geothermal or green hydrogen).

Ecological Footprint of Renewables

Although renewables emit far less over their lifecycle than fossil fuels, they still impose environmental costs that are hidden when only counting installed capacity.

Land Use and Habitat Disruption

Large solar farms and onshore wind projects demand vast land areas, often in fragile ecosystems. Wind farms can fragment habitats for birds and bats, while solar farms may displace native vegetation and disrupt wildlife corridors.

Resource Extraction for Materials

Manufacturing solar panels, wind turbines, and batteries requires rare earth elements, lithium, cobalt, and copper. Mining and refining these resources—often concentrated in the Global South—can cause water pollution, deforestation, and serious social impacts, including child labor.

End-of-Life Disposal and Recycling Challenges

Solar panels and turbine blades have limited lifespans and are difficult to recycle. Beginning in the 2030s, a surge of solar waste is expected, raising pressing questions about how to build a circular materials economy.

Why It Matters in Sustainability Thinking

Focusing solely on installed capacity obscures the actual energy delivered, the resilience of the grid, and ecological trade-offs. It risks creating misplaced incentives—for instance, maximizing large solar projects in ecologically sensitive zones instead of encouraging rooftop solar paired with conservation. More importantly, it misses the systemic picture: a truly sustainable energy system must be not only renewable but also **just, circular, resilient, and place-aware**.

What Should Be Tracked Instead (or Alongside Capacity)

To create a more accurate and holistic measure of renewable energy progress, metrics should include:

- capacity factor (actual output versus theoretical maximum)

- renewable penetration on the grid (percentage of electricity delivered)
- energy storage capacity
- land use intensity (MW per hectare)
- embedded emissions per kilowatt-hour
- materials circularity and recycling rates
- ecological and social impact assessments

Appendix XIII

What the "Uber Model" Implies in Energy Context

Just as **Uber** connects individual drivers (decentralized providers) with riders (demand) via a **real-time, platform-driven marketplace**, the **Uberized Energy model** suggests:

Table XIII.1 – Comparison: Uber Model of Transport vs. Energy

Uber Model (Transport)	**Uberized Energy Model**
Cars = distributed supply (drivers)	Rooftop solar, EV batteries, home generators
Riders = demand	Households, EV charging, small businesses
Platform = dynamic matching, optimization	Smart grid + AI + blockchain for load balancing
Surge pricing = real-time value signals	Time-of-use pricing / peer-to-peer energy trading
Idle time = underused resources	Unused battery storage, excess solar generation

Key Components of "Uberized" Energy

<u>Peer-to-Peer Energy Trading</u>

- Solar panel owners sell excess energy to nearby users.
- Enabled by blockchain smart contracts.

<u>Prosumers, Not Just Consumers</u>

- Each household or business can **produce**, **store**, and **sell** power.

<u>Real-Time Demand-Response</u>

- Smart appliances adjust based on grid signals or dynamic pricing.
- Microgrids respond autonomously to load changes.

<u>Flexible, Modular Infrastructure</u>

- Energy is no longer "centrally dispatched" only.
- Resilience built through distributed generation (solar rooftops, wind, biomass).

Analogous Features to Uber

- **Platform-Mediated Matching** – Just as Uber matches supply/demand of rides, energy platforms match power availability to local needs.
- **Decentralization** – No reliance on a single utility for generation—power flows from many sources.
- **Data-Driven Optimization** – Algorithms route power to highest-need areas, similar to how Uber dispatches cars.

Why This Matters

The Uber analogy helps reimagine power systems:

- from **pipeline thinking** (central → local)
- to **ecosystem thinking** (local ↔ local ↔ central)
- encouraging citizen participation, resilience, and real-time market efficiencies

Appendix XIV

Real World Examples of Entangled Interventions

Entanglement means that elements of a system are **not independent**; their outcomes are **correlated across space, time, and scale.**

Here are examples of – (i) **entangled, nonlinear, multi-lever interventions** that transcend traditional policy silos and (ii) its interpretation considering the Meadows *Leverage Points* (reproduced below from Chapter 5).

Meadows ***Leverage Points: Places to Intervene in a System***

12 leverage points ranging or levers from shallow—least powerful (adjusting parameters) to deep—most powerful (shifting paradigms).

1. **Constants, parameters, numbers** (e.g., subsidies, taxes, standards).
2. **Sizes of buffers** (relative to their flows).
3. **Structure of material stocks and flows.**
4. **Lengths of delays.**
5. **Strength of negative feedback loops, relative to the impacts they are trying to correct against.**
6. **Gain around driving positive feedback loops.**
7. **Structure of information flows.**
8. **Rules of the system** (e.g., incentives, punishments).
9. **Power to add, change, or self-organize system structure.**
10. **Goals of the system.**
11. **Mindset or paradigm** out of which the system arises.
12. **Power to transcend paradigms.**

Example 1: Carbon Pricing

Adjusting the "price" of carbon (a parameter) is a lever (#1), but:

- it affects **Energy Investment** (#3)
- alters **Consumer Behavior** (#5)
- shifts societal **Perception of Responsibility** (#10)
- interacts with global **Trade Systems** (#7)
- triggers backlash or momentum based on **Narrative Framing** (#6, #11)

Carbon pricing is not just a price mechanism or economic tool. It's an **entangled signal** in a multidimensional ecosystem reshaping culture, markets, and politics.

Example 2: Clean Manufacturing Mandates

Mandating industrial decarbonization (lever #5) does more than reduce emissions in factories:

- reshapes **Energy Demand Curves** (#6) by shifting industrial loads toward cleaner power;
- spurs **Green Innovation** (#4), unlocking new market niches and supply chains;
- alters **Trade Competitiveness** (#7) as global supply chains adapt to new compliance standards;
- impacts **Labor Markets** (#3) by requiring reskilling in clean production methods;
- influences **Narratives of Industrial Responsibility** (#2), potentially sparking public support or backlash.

Clean manufacturing mandates are not just a compliance tool; they are structural interventions that ripple through innovation ecosystems, geopolitics, and cultural perceptions of industry.

Example 3: Carbon Drawdown + Soil Health + Farmer Livelihoods

Integrating soil carbon capture with rural livelihoods is more than a climate solution:

- alters **Material Flows** (#6) through improved soil-water-carbon cycling;
- shifts **Agricultural Goals** (#3) from yield maximization to regenerative outcomes;
- reorients **Mindsets** (#2) from extractive agriculture to stewardship-based systems;
- creates **Economic Buffers** (#1) by diversifying farmer income streams via carbon credits.

This is an ecological, economic, and cultural transformation that redefines agriculture's role in climate and food systems.

Example 4: CRISPR-Edited Low-Lignin Trees + Pulping Efficiency + Forest Biodiversity

Biotech innovation in forestry affects more than paper mills:

- restructures **Biomass Efficiency** (#6) by lowering processing energy requirements;
- enables **System Evolution** (#4) by embedding biotech into sustainable forestry frameworks;
- expands **Industrial Goals** (#3) to include biodiversity and ecosystem services.
- shifts **Global Trade Flows** (#7) as bio-based materials compete with fossil-based inputs.

Low-lignin trees become a lever that connects genetics, manufacturing efficiency, biodiversity, and trade competitiveness.

Example 5: Rainwater Harvesting + Urban Heat Islands + Cultural Reclamation

Water harvesting is more than a climate adaptation measure:

- reinforces **Cooling-Resilience Loops** (#7) where green spaces reduce heat and encourage water retention;
- changes **Urban Planning Rules** (#5) through building codes and infrastructure retrofits;
- revives **Cultural Paradigms** (#2) by reclaiming indigenous water stewardship practices;

- alters **Urban Energy Use** (#6) as cooler microclimates reduce AC demand.

It's a convergence of environmental adaptation, urban design, and cultural renewal.

Example 6: Mindfulness Education + Climate Resilience + Intergenerational Ethics

Education that develops inner skills is a climate lever:

- reframes **Mindsets** (#2) toward empathy, foresight, and responsibility;
- enables **Paradigm Transcendence** (#11) by linking inner transformation to systemic reimagination;
- modifies **Institutional Parameters** (#1) through curriculum design and teacher training;
- strengthens **Community Resilience** (#7) by fostering collaborative and ethical decision-making.

Mindfulness is not a soft skill—it's a resilience infrastructure.

Example 7: EV Charging + Grid Flexibility + Job Access

Vehicle electrification isn't just mobility—it's infrastructure:

- reconfigures **Material Flows** (#6) by making EV batteries distributed storage assets;
- creates **Reinforcing Feedbacks** (#7) as bi-directional charging stabilizes the grid;
- improves **Information Flows** (#10) with real-time pricing and equitable access platforms;
- expands **Economic Goals** (#3) to include job creation in clean transport ecosystems.

It's a socio-technical transformation of mobility, labor, and energy.

Example 8: Storytelling + Policy Feedback + Narrative Legitimacy

Narratives shape systems as much as infrastructure:

- reorients **Paradigms** (#2) about belonging, responsibility, and justice;
- alters **Information Flows** (#10) by deciding whose voices influence policy;
- strengthens **Cultural Buffers** (#11) that protect societies from climate fatigue.

Storytelling is governance—it legitimizes action and sustains public will.

Example 9: Degrowth + Cooperative Economics + Ecological Restoration

Scaling back consumption is an active systems choice:

- changes **System Goals** (#3) from GDP growth to well-being;
- resets **Paradigms** (#2) toward sufficiency over excess;
- evolves **System Structures** (#4) toward cooperative, circular economies.

Degrowth is not austerity—it's design for planetary balance.

Example 10: Satellite Monitoring + Indigenous Land Maps + Legal Empowerment

Combining tech and traditional knowledge changes governance:

- improves **Information Flows** (#10) with real-time monitoring linked to stewardship;
- changes **Rules of the System** (#5) through land tenure reforms;
- reframes **Paradigms** (#2) by integrating indigenous and scientific epistemologies.

This is surveillance repurposed for sovereignty.

Example 11: Satellite Monitoring + Indigenous Land Maps + Legal Empowerment

Integrating satellite imagery with indigenous land mapping is more than a technical upgrade:

- enhances **Information Flows** (#10) by linking space-based monitoring directly to on-the-ground stewardship;
- changes **Rules of the System** (#5) through recognition of indigenous land rights and reforming tenure laws;
- shifts **Paradigms** (#2) by integrating Western geospatial technologies with indigenous ecological knowledge;
- creates **Positive Feedback Loops** (#7) as stronger legal protection enables more sustainable land management, which in turn reinforces biodiversity and carbon storage.

This is not just observation—it's empowerment through co-governance.

Table XIV.1 – Summary: High-Leverage Point Heatmap

Meadows #	**Description**	**Targeted In Examples**
#12	Transcending paradigms	6, 8
#11	Changing mindset/paradigm	3, 5, 8, 10, 11
#10	Changing the goal of the system	3, 7, 9
#9	Changing self-organization / Power to evolve system structures	4, 9
#8	Changing the rules of the system	2, 5, 10, 11
#7	Changing physical structures/material flows	2, 3, 4, 7
#6	Reinforcing positive feedback loops	5, 7, 11
#5	Negative feedback loops	
#4	Changing information flows	7, 8, 10, 11
#3	Buffers, constants, parameters	2, 3, 6
#2	Mindset or paradigm	3, 5, 6, 8, 9, 10, 11
#1	Constants & parameters	6

Appendix XV

The Quantum Change Spiral

Overview

The Quantum Change Spiral is a holistic framework that describes how transformative change emerges not as a step-by-step process but through resonance across multiple interconnected fields. Unlike linear models that assume predictable causality, the spiral reflects how identity, perception, behavior, collaboration, and amplification interweave to generate systemic shifts. Each field is distinct yet permeable, with feedback loops reinforcing the whole.

The Five Interlinked Fields

Being: Who We Are in the System

Change begins with self-awareness. *Being* encompasses identity, purpose, emotional literacy, and relational awareness. It acknowledges that sustainability transitions are not just technical but deeply personal. When individuals and institutions shift their inner narratives—from extractive logics to regenerative mindsets—every other layer of transformation becomes possible.

Example: Embedding Indigenous principles of stewardship into institutional values demonstrates how inner narratives shape outward practices.

Seeing: Mapping What Truly Is

This field emphasizes systems literacy—recognizing patterns, feedback loops, interdependencies, and thresholds. It draws from tools like foresight, scenario planning, and network analysis to reveal the invisible dynamics shaping outcomes.

Example: Climate tipping point modeling that highlights

nonlinear risks (ice melt, permafrost thaw) demonstrates the necessity of "seeing" beyond surface-level metrics.

<u>Acting: Aligning Behavior with Values</u>

Acting is about prototyping new ways of living and working, guided by regenerative principles. It involves small-scale experiments, behavioral nudges, and organizational practices that embody sustainability, allowing iteration and rapid learning.

Example: Local agrovoltaics initiatives that couple food and energy security exemplify aligned, place-based action.

<u>Weaving: Building Collective Intelligence</u>

Change is not achieved in isolation. *Weaving* refers to collaboration, co-creation, and polycentric governance. It stresses partnership across scales—communities, institutions, sectors—and emphasizes open-source learning to ensure that collective intelligence expands.

Example: The Costa Rican reforestation effort succeeded by weaving government policy with community-led stewardship.

<u>Amplifying: Spreading Signals Across Systems</u>

The final field addresses scaling: spreading innovations and narratives through digital infrastructures, cultural storytelling, and indices that synchronize markets with planetary thresholds. Amplification ensures that localized innovations resonate globally.

Example: A quantum carbon index, powered by AI and IoT, can harmonize fragmented efforts into coordinated global momentum.

Spiral Dynamics, Not Steps

The metaphor of a spiral emphasizes resonance and iteration rather than sequence. Fields overlap and reinforce one another. For example, shifts in *Being* influence how we See, which shapes *Acting*, while *Weaving* ensures these actions are collectively scaled through *Amplifying*. Change emerges from the interplay, not from completing stages in order.

Appendix XVI

Bio-Based Material Flow Schematics

Material Flow Diagram: [Forestry Waste → Dissolution in Solvent → Fiber Regeneration → High-Performance Textiles]

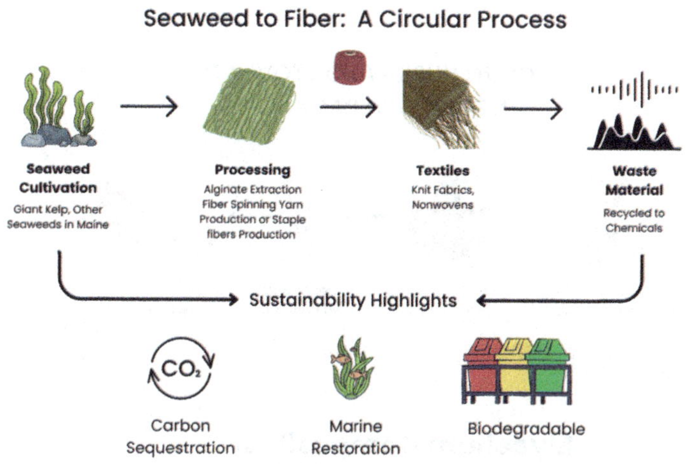

Figure XVI.1

Figure XVI.2

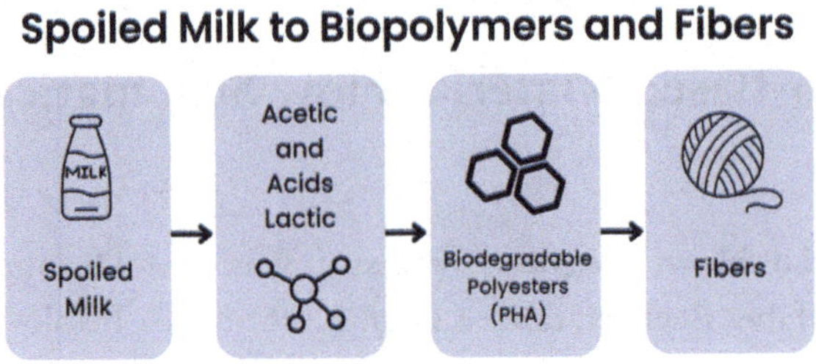

Figure XVI.3

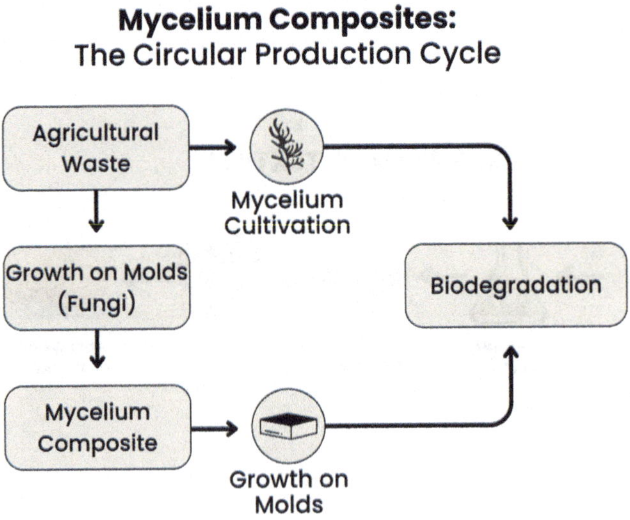

Figure XVI.4

Appendix XVII
The Sustainability Coherence Index (SCI)

Why We Need a New Metric

We live in a world where what we measure, shapes what we do. As outlined in Chapter 3, traditional sustainability metrics such as carbon emissions, GDP growth, or energy efficiency, while useful, often collapse the complexity of sustainability into a single observable. This reductionism can lead to distorted decision-making, unintended consequences, and blind spots in policy and innovation. The Sustainability Coherence Index (SCI) is proposed as an integrative, multidimensional alternative that better reflects the entangled nature of sustainability.

What Is the Sustainability Coherence Index (SCI)?

The SCI is a composite metric designed to evaluate the **coherence, integrity, and systemic soundness** of a policy, product, business model, community initiative, or urban plan from a sustainability perspective. Rather than isolate carbon or efficiency, the SCI evaluates the alignment of interventions with six domains: planetary boundaries, social equity, ecological resilience, circularity & regeneration, innovation quality, and cultural relevance.

Each domain can be scored from 0 to 100 based on a transparent indicator set and weighted per context (e.g., ecological vs. urban vs. corporate application).

Table XVII.1 – Six Domains of Coherence

Domain	Key Indicators	Systemic Rationale
Ecological Integrity	Biodiversity, native species protection, soil health	Sustainability starts with the land. Biodiversity and soil quality are keystone indicators of regenerative capacity.
Planetary Boundaries	GHG emissions, nitrogen/phosphorus cycles, land and water stress	Measures alignment with scientifically established limits to Earth's carrying capacity.
Circularity & Regeneration	Reuse rates, end-of-life design, material input/output ratios	Tracks systemic shift from linear to circular resource flows.
Social Equity	Access to clean energy, affordable housing, mobility, and fair wages	Sustainability without justice is incoherent. This domain tracks inclusion and redistribution.
Innovation Health	Local adaptive solutions, frugal innovation, open-source technologies	Encourages decentralization, creativity, and resilience in solutions.
Meaning & Engagement	Community participation, cultural coherence, mental wellbeing	Measures whether the intervention resonates with lived experience and collective purpose.

Aligning Evaluation Frameworks

Throughout this book, several frameworks are used to assess sustainability projects and interventions—ranging from energy–food–water linkages to carbon–justice structures, to broader socio-ecological categories. While these appear diverse, they are not fragmented; they are **nested within a single cohesive anchor: the SCI framework** outlined here. By treating SCI as the integrative compass, we can reconcile these domain-specific perspectives into a common evaluative architecture. This coherence ensures that ecological integrity, planetary boundaries, regeneration, social equity, innovation vitality, and meaning remain interconnected dimensions rather than isolated silos. Ending with SCI reinforces the book's central claim: that sustainability requires not linear optimization of parts, but a **coherent orchestration of whole-system health**.

Table XVII.2 provides a crosswalk that shows how all frameworks—those introduced in Chapters 3, 4, 5, 6, and 10—map back onto the six SCI dimensions: ecological integrity, planetary boundaries, circularity and regeneration, social equity, innovation health, and meaning/engagement. This harmonization ensures coherence while preserving nuance. Domain-specific categories (e.g., mobility, finance, culture) are shown as **contextual applications** of the broader SCI framework.

Table XVII.2 – Alignment of Evaluation Frameworks with the SCI Framework

Framework / Location	Categories Used	Alignment with SCI Dimensions	Notes
SCI (Appendix)	Ecological integrity, planetary boundaries, circularity & regeneration, social equity, innovation health, meaning & engagement	Anchor framework	Master structure; all other frameworks map here.
Earth's Climate: The Archetype of a CAS (Chapter 3)	Energy, food, water, biodiversity, economy, equity, institutions	Ecological integrity, planetary boundaries, social equity, institutions ↔ meaning/engagement	Resource and governance domains framed as systemic levers.
Transition to Application (Chapter 4)	Energy, food, water, finance, culture	Ecological integrity, planetary boundaries, social equity, finance ↔ innovation health, culture ↔ meaning/engagement	Illustrates multi-sector integration; adds finance & cultural dimensions.
The Legacy of Donella Meadows (Chapter 5)	Ecological integrity, planetary boundaries, circularity, social equity, innovation health	Direct overlap with SCI (except meaning/engagement)	Considered an early prototype of SCI.
Superposition in Strategy: Designing for Multiple Outcomes (Chapter 6)	Carbon, water, energy, food, equity	Carbon, water, energy, food ↔ planetary boundaries, equity ↔ social equity	Operational metrics for sector-specific tracking.
Entangled Energy System (Chapter 10)	Water, food, mobility, carbon, equity	Planetary boundaries, social equity, mobility ↔ meaning/engagement	Scenario lens with emphasis on urban/transport systems.

By consolidating these into one visual guide, readers can see how the different parts of the book interlock, how thematic emphases (like food, water, mobility, or equity) sit within the SCI structure, and how the framework can be practically applied to diverse contexts without fragmenting the core argument.

Visual Format: The SCI Dashboard

SCI can be visualized as a **radar plot** or **hexagonal coherence wheel**, where each spoke represents one domain. A high-SCI intervention would display a balanced, outward-reaching profile across domains. A low-SCI score may have strong emissions reduction but low equity or poor ecological coherence. An example is presented in Table XVII.3 below.

Table XVII.3 – Sample Application: A Rooftop Solar Program

Domain	Evaluation
Ecological Integrity	High, if no forest cleared or e-waste created
Planetary Boundaries	Medium-High, strong decarbonization benefits
Circularity	Medium, depends on panel reuse or recycling infrastructure
Social Equity	Variable: High if access expanded for low-income communities; low if it reinforces energy elitism
Innovation Health	Medium: Technological standard but can be locally manufactured
Meaning & Engagement	High if community co-ownership; low if purely corporate-driven

Composite SCI Score: 78/100 (context-dependent)

From Score to Story

Unlike single metrics, the SCI offers a **narrative dashboard**: Where is the intervention strong? Where is it incoherent? What are the tradeoffs? What feedbacks are being amplified or suppressed?

Figure XVII.1

Theoretical Foundations

The SCI draws inspiration from:

- **Donella Meadows' leverage points**, especially on shifting mindsets and system design;
- **The Planetary Boundaries framework** (Rockström et al., 2009);
- **The Doughnut Economics model** (Raworth, 2017);
- **Human-scale development** and post-growth economics.

Implications for Practice

- **For Policy** – Enables governments to compare programs not just on efficiency but on coherence and justice.
- **For Business** – Offers ESG+ metrics aligned with long-term resilience, not just optics.

- **For Finance** – Can inform climate finance flows toward interventions with high systemic leverage.
- **For Communities** – Can support participatory decision-making and values-based planning.

Final Note: A Metric with Humility

The SCI is not meant to replace all metrics, but to **complement** them. It asks different questions, such as not just "How much CO_2?," but also "How just? How regenerative? How coherent?"

In a nonlinear world, coherence is power. The SCI aims to be a compass, not a speedometer—pointing toward systems that can survive, adapt, and thrive. It introduces a multidimensional, systems-aligned metric to assess interventions beyond carbon, grounded in ecological, social, and narrative coherence.

Bridging Paragraph for Conclusion

Across the book, multiple lenses—energy–food–water systems, carbon–justice tradeoffs, ecological thresholds, and innovation metrics—have been introduced to analyze sustainability transitions. While these appear diverse, they are not fragmented; they are **nested within the SCI framework** outlined here. By treating SCI as the integrative compass, we can reconcile these domain-specific perspectives into a common evaluative architecture. This coherence ensures that ecological integrity, planetary boundaries, regeneration, social equity, innovation vitality, and meaning remain interconnected dimensions rather than isolated silos. Ending with SCI reinforces the book's central claim: that sustainability requires not linear optimization of parts, but a **coherent orchestration of whole-system health**.

Appendix XVIII

The Future of Climate Action

Nonlinear Pathways to Transformation

From Linearity to Living Systems

The climate crisis is not a single-variable problem to be solved; it is a living system that learns, reacts, and reorganizes. Traditional climate strategies—based on proportional inputs and predictable returns—misread this reality. Nonlinear dynamics remind us that **small interventions can create cascading transformations**, while inertia in one sector can destabilize entire networks.

The path forward demands *coherence across complexity*: connecting ecological integrity, equity, innovation, and finance within a shared framework. Progress must be measured not through isolated metrics but through how interventions interact—whether they amplify resilience or accelerate breakdowns. This systems coherence, rather than policy uniformity, is the foundation of nonlinear mitigation.

Understanding Nonlinearity in Climate Behavior

Climate systems behave as dynamic feedback networks, where cause and effect are rarely proportional or predictable.

- **Albedo feedback** – Melting Arctic ice darkens oceans, absorbing more heat, which accelerates further melt—a self-reinforcing spiral.
- **Permafrost thaw** – Trapped methane escapes, intensifying warming that deepens the thaw.
- **Coral bleaching** – Local stress events accumulate until entire reef ecosystems flip to degraded states.
- **Ocean currents** – Freshwater from Greenland meltwater can stall circulation, altering rainfall and heat distribution globally.

Each illustrates a pattern: **thresholds emerge gradually, then break suddenly**. Once crossed, reversal becomes exponentially harder.

Complexity as a Design Principle for Mitigation

Complexity is not a barrier to climate action—it is the **blueprint** for designing systems that can adapt and endure. In a nonlinear context, effective mitigation follows three design logics:

1. **Diversity and redundancy**: Multiple energy sources, supply chains, and governance layers act as buffers against collapse.
2. **Feedback fluency**: Continuous sensing, data loops, and real-time learning allow early correction before tipping points.
3. **Distributed intelligence**: Empowering communities, cities, and enterprises to act autonomously within shared guardrails enables adaptive coherence without central command.

Complexity-driven mitigation converts fragility into *antifragility*: each disturbance becomes an opportunity for innovation.

Governance in a Nonlinear Era

Linear governance models assume predictability; nonlinear governance anticipates surprise. Future climate institutions must behave like adaptive organisms—**learning, recalibrating, and evolving** through continuous interaction with environmental and social signals.

Policy design under nonlinearity should:

- Embed *experimentation* through pilot zones and dynamic feedback budgets.
- Institutionalize *adaptive reviews*—rolling policy cycles similar to the Dutch Delta Program's iterative water governance.
- Couple *mitigation and adaptation* so that resilience measures (like forest restoration or microgrids) double as emissions reduction tools.
- Integrate *tipping-point intelligence*—monitoring indicators (ice-sheet melt rate, carbon sink saturation, food system volatility) to adjust course proactively.

From Complexity to Coherence

Fragmentation—between ministries, frameworks, or metrics—is itself a nonlinear risk. The next generation of climate governance must **translate multiple worldviews into one coherent compass.**

Your proposed **Sustainability Coherence Index (SCI)** offers precisely this: a way to align sectoral metrics (water, energy, circularity, justice, innovation) into a common evaluative language. Such coherence transforms data into design—enabling policymakers, financiers, and citizens to see how their decisions interlock across domains.

Nonlinear Case Studies of Transition

- **The Dutch Delta Program** – Adaptive flood management cycles every six years, combining physical infrastructure with policy learning—an exemplar of iterative, reflexive governance.
- **Costa Rica's Forest Transition** – Incentives for reforestation created a positive feedback loop—policy, community, and biodiversity reinforcing each other until forest cover exceeded 50%.
- **Cape Town's Day Zero Recovery** – Dynamic water governance turned crisis into innovation—real-time feedback, behavioral shifts, and decentralized planning built resilience faster than linear infrastructure investment.
- **Urban energy microgrids in India** – Localized solar + storage networks illustrate distributed nonlinearity—small, modular actions summing to systemic transformation.

Each shows that resilience is not static robustness but *the capacity to reorganize faster than disruption unfolds.*

Nonlinear Planning Tools and Methods

Future-ready mitigation will combine scientific modeling with participatory intelligence. Emerging tools include:

- **complex systems simulation** integrating feedback loops and multiple equilibria;

- **ensemble modeling** for uncertainty management;
- **scenario planning** that explores plausible discontinuities, not just smooth trajectories;
- **AI and machine learning** to detect early-warning indicators of instability;
- **dynamic adaptive pathways**—planning structures that map multiple futures and pivot options.

These methods move climate planning from "forecasting and fixing" to "learning and steering."

Evaluating Climate Action Under Nonlinearity

Table XVIII.1 – Summary

Domain	Evaluation Focus	Interpretation
Ecological Integrity	Forest protection, no net deforestation, biodiversity health	Sustained resilience of natural carbon sinks
Planetary Boundaries	Decarbonization rate and materials footprint	Staying within safe operating space
Circularity	Product reuse, recycling, modular design	Resource feedbacks reinforcing sustainability
Social Equity	Access to green energy, fair transitions	Inclusion reduces systemic vulnerability
Innovation Health	Local manufacturing, adaptive technologies	Enhances regional self-sufficiency and learning

The purpose of evaluation is not only measurement—it is **alignment**. Metrics reveal whether the system learns coherently or fragments under stress.

Nonlinear Solutions for a Dynamic Century

Climate mitigation in this century must function like a living network—distributed, diverse, and self-correcting. Key pathways include:

- **ecosystem regeneration** as a dual carbon and resilience engine;
- **modular renewable systems** that can scale or isolate during disruptions;
- **circular bioeconomy transitions** that transform waste into resource loops;
- **social innovation accelerators** that replicate successful behaviors (e.g., low-carbon diets, sharing economies);
- **adaptive finance**—green bonds or insurance instruments that recalibrate to risk in real time.

Corporations like *Unilever* and *Ørsted* are already embedding feedback-based adaptation into supply chains and investment cycles—early steps toward nonlinear governance.

The Shift Ahead: From Incrementalism to Systemic Regeneration

The future of climate action will be defined not by carbon metrics alone but by how **fast and fluidly systems can learn**. Nonlinear mitigation reframes climate change as a problem of *coherence and adaptation*, not simply of emissions. It requires humility—the recognition that control is an illusion—and courage: to design policies, technologies, and cultures that can pivot amid surprise.

When complexity is embraced as an ally, not an obstacle, mitigation becomes *evolutionary* rather than *reactive*. Humanity's task is no longer to "stabilize the climate" but to **co-evolve** with it—cultivating regenerative systems capable of thriving in motion.

Key Takeaway

The Future of Climate Action: From Linearity to Living Systems

Figure XVIII.1 summarizes the key takeaway well. The climate system teaches that the future is neither linear nor controllable. Yet it is **navigable** through coherence, feedback, and resilience intelligence. The era ahead belongs to those who design policies, technologies, and cultures that *learn faster than the crisis unfolds*—transforming climate action from damage control into the art of systemic regeneration.

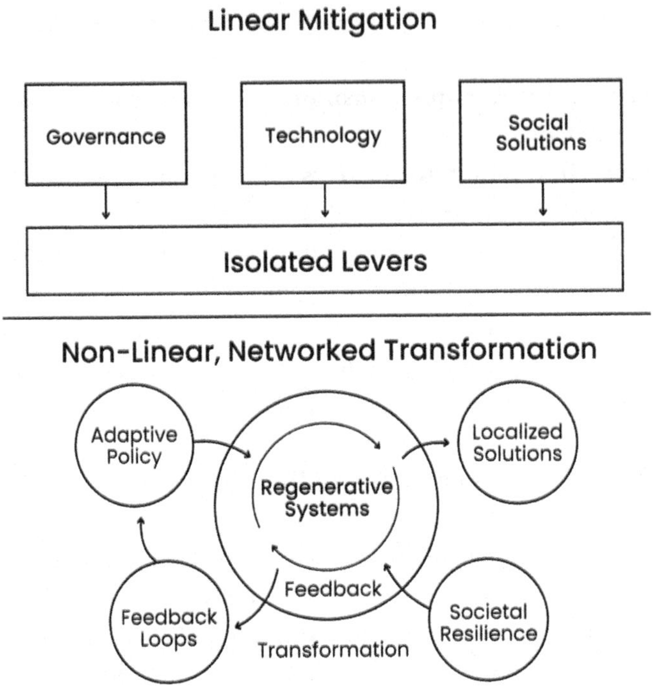

Figure XVIII.1

Appendix XIX
Nonlinear Causality

Reinforcing (Self-Amplifying) Feedbacks — More Depth

The intellectual lineage of nonlinear causality predates its formal scientific articulation in the 20th century. Early ecosystem thinkers like Alexander von Humboldt emphasized how human interventions could trigger cascading, multi-dimensional consequences for land, water, and climate. His holistic worldview foreshadowed today's complexity science and reminds us that "quantum sustainability" is not entirely new, but a return to seeing nature as an entangled system.

Why Nonlinear Causality Matters

Most climate and ecological systems are nonlinear, meaning cause and effect are not proportional. Small actions can unleash massive changes—positive or negative—while large interventions may yield minimal results. This is why climate mitigation strategies must always consider feedback loops and tipping points: without them, projections risk being dangerously optimistic or excessively conservative.

Reinforcing (Self-Amplifying) Feedbacks

Reinforcing loops accelerate change once momentum starts. Examples include:

- **Technology learning curves** – Every doubling of solar or battery output reduces costs, spurring adoption and lowering costs further. Policy levers like subsidies, public procurement, and standards amplify this cycle.
- **Urban mobility transitions** – More bike lanes increase cyclists,

which strengthens political support for further lanes; similarly, denser EV charging infrastructure accelerates EV uptake and private investment.

- **Deforestation–rainfall loop** – Forest loss reduces evapotranspiration, lowering rainfall, raising fire risk, and driving further forest loss (as in the Amazon and Congo).
- **Narratives and markets** – Positive climate narratives attract capital, scale clean technologies, reduce costs, and reinforce the narrative.
- **Water vapor and ice–albedo feedbacks** – Warming increases evaporation and greenhouse gas effect of water vapor; melting ice reduces reflectivity, accelerating warming.
- **Urban heat island intensification** – Rising temperatures drive more air-conditioning, which generates waste heat and further warming.

Design cue: Identify small early pushes that trigger virtuous cycles, and stack policies across cost, convenience, and culture to cross thresholds simultaneously.

Dampening (Self-Limiting) Feedbacks

Dampening loops slow or stabilize trajectories. For example, vegetation regrowth draws down CO_2 after disturbance, though limited by water, nutrients, and heat stress. Volcanic aerosols temporarily cool the planet by reflecting sunlight. Economic saturation slows market adoption once growth limits are reached. Renewable "cannibalization" occurs when abundant solar or wind depresses prices, discouraging new builds unless storage or flexibility grows in parallel.

Design cue: When reinforcing loops are triggered, anticipate and co-manage the dampening effects—for instance, scale storage as solar expands.

Thresholds, Tipping Points, and Hysteresis

Thresholds represent irreversible state shifts; hysteresis means recovery requires more than reversing the original driver. Examples include:

- a slowdown of the Atlantic Meridional Overturning Circulation (AMOC) reshaping weather systems;
- permafrost thaw releasing methane pulses;
- coral bleaching events causing irreversible reef mortality;
- and lakes flipping suddenly into eutrophication, requiring deeper nutrient cuts to restore.

Design cue: Build precautionary buffers and act before thresholds are crossed.

<u>Early-Warning Signals and Diagnostics</u>

Systems approaching tipping points often show signs such as slower recovery from small disturbances, increasing variance in time-series data, flickering between states (e.g., clear and turbid lakes), or network fragility concentrated in key nodes. These signals should be operationalized in **Dynamic Adaptive Policy Pathways (DAPP) or Robust Decision-Making (RDM)** frameworks to trigger timely policy pivots.

<u>Cross-Sector Synergies and Co-Evolution</u>

Nonlinear gains emerge when sectors evolve together: EV battery scale reduces storage costs for grids; green hydrogen pilots in fertilizers and refining unlock applications in shipping and steel; heat pumps paired with smart tariffs shift demand to renewable-rich hours; smart grids combined with electrified transport create demand-response capabilities.

Design cue: Favor parallel transitions with shared infrastructure and pricing signals that reward flexibility.

<u>Policy Design Heuristics</u>

Policies should harness reinforcing loops while guarding against risks. Practical rules of thumb include:

- prime the pump with time-limited deployment support;
- co-develop complements (storage, grids, flexibility) to prevent cannibalization;
- treat policies as hypotheses—build in review and iteration;
- budget for buffers in critical systems like water or grids;

- guard against lock-in by avoiding long-lived high-carbon assets;
- target leverage points such as procurement standards;
- diversify pilots and scale the winners.

Positive Tipping Points

Positive transitions can be triggered through mandates and standards (e.g., zero-emission vehicles, clean power codes), making sustainable choices more convenient than alternatives, leveraging social contagion through visible adoption (rooftop solar, e-bikes), and sending finance signals such as green mortgages or carbon-aware payment tools (linked to the eCarbon Card in Chapter 11).

Design cue: Stack cost, convenience, culture, and capital to tip systems rapidly.

Managing "Limits to Success"

As transitions succeed, bottlenecks arise: grid congestion and permitting delays, workforce skill shortages, or critical mineral constraints. Proactive investment in corridors, training programs, recycling, substitution, and efficiency reduces these risks.

Mini-Vignettes Across Domains

- Urban cooling through trees, cool roofs, and water features reduces AC demand and waste heat.
- Adaptive fisheries quotas prevent stock collapse and avoid hysteresis.
- Rewetting peatlands flips them from carbon sources to sinks.
- Regenerative agriculture builds soil carbon and water retention, tipping adoption once profitable and socially normalized.
- In fashion, improved recycled fibers, extended producer responsibility, and labeling can trigger fast nonlinear take-back loops.

Metrics for Nonlinearity

Single KPIs are insufficient. Resilience requires metrics such as diversity and redundancy indices, flexibility minutes (responsive load per peak MW), carbon abatement with co-benefits, network potential value (stability contributions beyond ROI), and leading-indicator dashboards (variance, autocorrelation, signposts).

Key Takeaway

Nonlinear causality is not just a scientific concept—it is a design principle for climate action. Small, well-placed interventions can trigger systemic transformations, while neglecting feedbacks risks stalling progress.

Appendix XX

Incorporating Feedback Loops and Nonlinear Dynamics into Sustainability Policymaking

Why Nonlinear Thinking Matters

As noted in recent analyses such as *Carbon Brief's* discussion of feedback loops and nonlinear climate dynamics, environmental and social systems rarely behave in predictable, stepwise ways[137]. Small interventions can trigger cascading or irreversible transitions, while seemingly stable systems can reorganize abruptly once hidden thresholds are crossed. Recognizing these feedback-rich patterns is essential for designing **policies that learn, adapt, and evolve**, rather than simply control.

Traditional "predict-and-fix (or control)" models assume continuity, but sustainability governance now requires **anticipatory intelligence**—continuous sensing, iterative adjustment, and multi-level resilience. This is the essence of nonlinear policymaking: to expect surprises, not be undone by them.

Mapping and Understanding Feedbacks

Policymaking must begin with mapping feedback structures using causal loop diagrams, system dynamics modeling, and network analysis. This helps identify amplifying (positive) and stabilizing (negative) loops. By making tipping points visible, mapping also highlights where small interventions can yield disproportionate benefits—for instance, restoring wetlands to improve both flood control and carbon storage.

Adaptive Management

Interventions should be treated as evolving experiments rather than fixed solutions. Adaptive management involves continuous monitoring, evaluation, and adjustment. Regulations, incentives, and urban planning must remain flexible enough to adjust to technological change, ecological shifts, and social behavior without locking societies into rigid or outdated pathways.

Scenario Planning and Preparedness

Policies should not assume gradual linear change. Instead, scenarios must account for sudden threshold crossings, cascading failures, and abrupt regime shifts. Real-time monitoring systems—for biodiversity, water, or infrastructure stress—serve as early-warning tools, allowing preemptive action before nonlinear changes spiral out of control.

Building System Resilience

Efficiency alone often erodes buffers like redundancy and diversity, which are essential for resilience. Stronger systems diversify supply chains, restore degraded ecosystems, build community-level adaptation, and encourage cross-sector collaboration. These measures allow systems to absorb disturbances and reorganize effectively.

High-Leverage Points

Identifying and acting on critical nodes is key. Social norms, keystone ecosystems, and infrastructure chokepoints represent points where small, targeted actions can cascade outward. Strategic interventions include behavior-change campaigns, renewable energy subsidies, or habitat protection—each capable of stabilizing larger systems.

Advanced Modeling and Analytics

Agent-based models, system dynamics, and Bayesian networks can simulate outcomes in feedback-rich environments. Machine learning and big data analytics help detect early-warning signals, threshold behavior, and hidden correlations, improving foresight and responsiveness.

Stakeholder Engagement and Participatory Approaches

Inclusive policymaking strengthens both effectiveness and legitimacy. Communities, Indigenous groups, scientists, businesses, and NGOs can help map feedbacks and co-create adaptive strategies. Participatory scenario planning integrates diverse knowledge systems, values, and perspectives, improving policy alignment with local realities.

Examples in Practice

- **Urban Water Management** – Shifting from fixed water-use rules to adaptive, real-time demand management based on rainfall and consumption feedbacks.
- **Corporate Sustainability** – Unilever's soil health and community resilience programs reduce supply-chain risks from climate shocks.
- **Ecosystem Restoration** – Programs designed with early-warning triggers help prevent ecological tipping points.
- **Everglades Restoration** – Hydrological and ecosystem models guide large-scale restoration, explicitly incorporating nonlinear dynamics.
- **Baltic Sea Management** – Cross-border governance structures manage nutrient loading and resilience in a shared marine system.

Case Studies

- **Everglades (USA)** – A comprehensive restoration plan explicitly addressed nonlinear dynamics and uncertainty.
- **Baltic Sea** – Collaborative international management accounted for feedbacks and regime shifts, demonstrating the value of cross-border governance.

Role of Stakeholder Engagement

Collaborative governance and public participation are indispensable. Engaging stakeholders builds trust, enhances legitimacy, and improves resilience by integrating diverse perspectives into decision-making. The lesson is clear: managing nonlinearity requires governance that is as participatory and adaptive as the systems it seeks to influence.

Conclusion and Key Takeaway

Nonlinear dynamics are not peripheral—they are central to environmental and social systems. Effective sustainability policymaking requires shifting from static, control-oriented strategies to dynamic, adaptive governance. The most effective policies acknowledge feedback loops, prepare for uncertainty, and adapt continuously.

Embedding nonlinear thinking means cultivating humility in forecasting and ambition in resilience-building. By anticipating abrupt change and positioning interventions to work with—rather than be destabilized by—complexity, policymakers can catalyze systemic transformations and prevent destructive tipping points. The goal is not merely to manage complexity but to navigate it skillfully, turning uncertainty into a foundation for sustainable change.

Part VIII: Supplemental

Discussion Guide for Book Clubs, Classes, or Teams

These questions are designed to provoke rich, interdisciplinary conversations and to help readers personally integrate the book's concepts.

❖ Section I: Shifting Mindsets

1. How does a quantum metaphor change the way you think about climate change compared to a classical metaphor?
2. What is the difference between "solving a problem" and "shaping a system"? Where do you see yourself in that distinction?
3. Which traditional sustainability assumptions did the book challenge for you?

❖ Section II: Systems and Signals

4. What surprised you most about the idea of carbon as a **signal**, not just a pollutant?
5. If you could visualize one feedback loop in your daily life or organization, what would it be? What signal would you track?
6. What would it look like to build "coherence" into your city, workplace, or household?

❖ Section III: Leadership and Agency

7. How do you relate to the concept of "quantum leadership"? Where are you leading (or could lead) under uncertainty?
8. What is the difference between a "footprint" and an "imprint" in your context?
9. Which of the five spheres of quantum citizenship felt most actionable for you?

❖ Section IV: Visioning and Futures

10. In Chapter 17, which of the future archetypes (Regenerative Civilization, Cascading Crisis World, Quantum Society) felt most real or most urgent?
11. Have you ever done "backcasting" in your personal or professional life? How would this approach change your strategy today?

❖ Section V: Personal Integration

12. What is your **coherence field**? What values, beliefs, and actions do you try to align?
13. Where are you experiencing dissonance between your knowledge and your action?
14. What one story, metaphor, or visual from this book will stay with you the longest?
15. How can you support others in becoming quantum citizens—through mentorship, dialogue, or example?

Reflection Journal Prompts for Readers

These prompts are ideal for individual contemplation, journaling, or organizational retreats.

Weekly Integration Prompts

- This week, where did I act from clarity instead of reactivity?
- What **signal** did I emit through my choices, words, or tone?
- Where did I witness **entanglement**—interconnection I hadn't seen before?
- How did I respond to uncertainty—with paralysis, or with possibility?

Deep Inquiry Prompts

- What future am I most loyal to? What world am I willing to **walk back from**?
- What would it mean to shift from "doing sustainability" to "**being sustainable**"?
- What part of me feels **fragmented** or out of coherence with my values?
- What feedback loops do I ignore—and what would it take to listen to them?

Future Design Prompts

- Imagine it's 2040 and we've stabilized the climate under 2°C. What did we do right?
- What three innovations—technological, social, cultural—enabled the transition?
- What are the next **three small experiments** I can launch in my own life or work?

Mini Quantum Thought Experiments in Sustainability

Schrödinger's Forest

Thought Prompt

Is the forest sustainable if no one is measuring it?

In this paradox, a biodiverse forest in a remote biome is neither protected nor endangered—until a sustainability team arrives with satellite imaging, baseline data, and ESG metrics.

Before observation, its fate is indeterminate: it *may* be thriving or it *may* be degraded.

Once measured, the forest collapses into a state—**quantified, managed, and perhaps even exploited**.

Do metrics help us protect, or do they create the illusion of control?

The Entangled Emissions

Thought Prompt

Can a ton of CO_2 emitted in Mumbai be offset by a tree planted in Kenya?

Imagine two particles—carbon atoms—produced and absorbed in different geographies, timelines, and socioeconomic conditions.

Are they truly *equal*?

In a classical frame, they cancel. In a quantum lens, their **entanglement includes context:**

- equity
- local air pollution
- water tradeoffs
- community governance

Carbon is not just a molecule—it is a relationship.

The Quantum Refrigerator

Thought Prompt

If your fridge is smart enough to optimize energy use, but you still fill it with beef, is it sustainable? This paradox questions the **substitution of efficiency for sufficiency.**

We observe a system (the fridge), optimize it technologically, and declare success—ignoring upstream emissions, supply chains, and behavioral drivers.

In sustainability, efficiency is a local optimization; coherence is the global one.

Heisenberg's Policy Uncertainty

Thought Prompt

The more precisely we try to predict the effect of a policy, the less political will remains to act on it.

When policies are debated, their impact must be modeled. But:

- too much modeling = delayed action
- too little modeling = potential failure

The act of "knowing" (with certainty) often delays "doing."

The Superpositioned Consumer

Thought Prompt

Are you sustainable or extractive?

The same individual:

- buys organic food
- flies twice a month
- works in fossil finance
- volunteers for a rewilding NGO

Until observed (or forced to choose), they exist in a **superposition of sustainability states**—both responsible and complicit.

Identity collapses into coherence only when choices align.

The Quantum Regenerator

<u>Thought Prompt</u>

What is the smallest unit of regeneration?

- Is it a compost pile?
- A community garden?
- A change in worldview?
- Or a viral TikTok explaining circular economies?

In quantum terms, the regenerative signal is **not a scale—it is a frequency.**

The power of regeneration lies not in size, but in resonance.

The Carbon Tunnel

<u>Thought Prompt</u>

Can you "quantum tunnel" through institutional resistance?

In physics, quantum tunneling allows particles to leap through barriers they could not overcome classically.

Could the same apply to:

- A revolutionary climate narrative?
- An underfunded climate startup?
- A radical policy proposal?

With enough alignment, even dense systems become permeable.

The Feedback Ghost

<u>Thought Prompt</u>

What if feedback loops exist, but we no longer respond? The ocean warms. Coral dies. Fisheries collapse. But no policy is triggered. The feedback is visible—but no longer functional.

Systems collapse when feedback becomes ambient noise.

The Pedagogical Observer Effect

Thought Prompt

Does teaching sustainability change the learner—or the teacher?

In workshops, facilitators track behavior change. But often, it is the educator who transforms—challenged by questions, stories, discomfort.

Knowledge transfer is a two-way entanglement.

The Tipping Point Inversion

Thought Prompt

What if we just passed the threshold—and didn't notice? Some planetary tipping points (e.g., Amazon dieback, Arctic ice loss) may be **silent**—only obvious decades later.

By the time we confirm a collapse, the collapse has already become self-propagating.

Glossary of Systems, Quantum, and Sustainability Terms

Glossary is drawn from systems science, quantum mechanics (as metaphor and science), complexity theory, governance, and sustainability—covering terms used throughout the chapters and appendices,

Adaptive Capacity

The ability of a system—whether ecological, social, or technological—to adjust, self-organize, and evolve in response to shocks or changing conditions while retaining its essential functions.

Adaptive Management

A policy and planning approach that treats interventions as experiments, continuously learning from real-world feedback and adjusting strategies accordingly.

Agent-Based Modeling (ABM)

A simulation method that models the actions and interactions of autonomous agents (individuals, organizations, species) to assess their effects on the system as a whole.

Albedo

The fraction of incoming sunlight reflected by a surface. Ice and snow have high albedo; when they melt, darker surfaces absorb more heat, amplifying warming.

Amplifying Feedback (Positive Feedback)

A loop where an initial change triggers effects that reinforce or intensify that change, potentially leading to runaway processes (e.g., Arctic ice melt reducing reflectivity and accelerating warming).

Antifragility

A property of systems that gain strength or capacity from shocks, volatility, or disruption, rather than merely resisting damage.

Bayesian Networks

Probabilistic graphical models that represent variables and their conditional dependencies, useful for analyzing uncertain, complex systems.

Behavioral Economics

The study of how psychological, social, and emotional factors influence economic decisions, often deviating from "rational actor" assumptions.

Bifurcation

A point where a small change in system parameters causes a sudden qualitative shift in behavior or structure.

Bio-Based Materials

Materials derived from renewable biological resources such as plants, fungi, algae, or agricultural residues.

Biomimicry

The design of products, systems, or policies inspired by natural processes, structures, and strategies.

Braess's Paradox

A counterintuitive phenomenon in network theory where adding capacity (e.g., building more roads) can worsen congestion because of how users adapt their routes.

Carbon Index

A composite indicator that tracks carbon emissions, removals, or intensity across multiple sectors to serve as a market and policy signal.

Carbon Pricing

Assigning a monetary value to greenhouse gas emissions to internalize environmental costs and incentivize low-carbon solutions.

Cascading Effects

A chain of impacts triggered when a change in one part of a system leads to sequential disruptions in other parts. [A sequence of events where one change triggers a chain of interconnected responses across systems.]

Citizen Science

Research and data collection undertaken in whole or in part by members of the public, often contributing to large-scale monitoring and decision-making.

Circular Economy

An economic model that aims to eliminate waste and keep resources in use for as long as possible through reuse, repair, remanufacturing, and recycling.

Climate Tipping Point

A critical threshold in the climate system beyond which a small change can cause a large, often irreversible shift to a new state.

Coherence Without Uniformity

A principle in polycentric governance where actors share overarching goals but apply diverse strategies suited to their contexts.

Complex Adaptive System (CAS)

A dynamic system composed of interconnected agents that adapt and evolve in response to interactions with each other and the environment.

Complexity Science

The interdisciplinary study of systems with many interacting parts, where patterns emerge without centralized control.

Confirmation Bias

The human tendency to seek and interpret information in a way that confirms preexisting beliefs, potentially overlooking contradictory evidence.

Critical Node

A component of a system whose functioning disproportionately influences overall performance or resilience.

Decentralized Governance

Decision-making distributed across multiple levels and actors rather than concentrated in a single central authority.

Decoherence (Quantum)

The process by which quantum systems lose their coherent superposition states due to interaction with the environment; metaphorically, the breakdown of alignment in governance or networks.

Delayed Causality

A condition in which the effects of an action occur much later than the cause, making cause-and-effect relationships harder to detect.

Edge-Powered Networks

Decentralized digital or energy systems where computing or generation occurs at or near the source rather than centrally.

Emergence

The arising of complex patterns, properties, or behaviors that cannot be predicted solely from the sum of a system's parts or the properties of individual components.

Enabling Constraints

Rules or boundaries that guide system behavior toward beneficial outcomes without overly restricting diversity or innovation.

Entanglement (Quantum)

A phenomenon where particles share states across space so that actions on one affect the other; metaphorically, the deep interconnection of issues and systems.

Eutrophication

A process where excessive nutrient input (mainly nitrogen and phosphorus) fuels rapid algal growth in aquatic systems, leading to oxygen depletion, loss of biodiversity, and ecosystem collapse. Metaphorically, it illustrates how unchecked positive feedbacks—often driven by human activity—can tip stable systems into degradation.

Feedback Loop

A process in which a system's output feeds back into the system as input, either amplifying change (positive feedback) or stabilizing it (negative feedback).

Fractals

Structures that exhibit self-similarity across scales; metaphorically, patterns of governance, ecosystems, or networks repeating across different levels.

Governance Deficit

The mismatch between existing institutional capacities and the demands of managing complex, interdependent challenges.

Greenland Ice Sheet Melt Feedback

A positive feedback loop where melting reduces reflectivity (albedo), causing more heat absorption and further melting.

Hysteresis

The tendency of a system to remain in a changed state even after the initial cause of that change is removed.

Ice-Albedo Feedback

A self-reinforcing process where melting ice reduces reflectivity, causing further warming and more melting.

Interoperability

The capacity for different systems, platforms, or institutions to work together through shared protocols and standards.

Leverage Point

A place in a complex system where a small change can produce significant, lasting transformation.

Linearity Illusion

The mistaken belief that system outputs are proportional to inputs, ignoring threshold effects and feedback dynamics.

Living System

A self-organizing system—such as an ecosystem, community, or adaptive organization—that maintains and regenerates itself over time.

Nonlinear Dynamics

System behavior where cause and effect are not proportional, often producing abrupt changes or multiple possible stable states.

Methane Release Feedback

A potent warming loop where thawing permafrost releases methane, which accelerates further warming and thawing.

Nestedness

The arrangement of systems within systems, where each level interacts with and influences others.

Nonlinear Dynamics

System behavior where outputs are not directly proportional to inputs, often producing thresholds, sudden shifts, or multiple stable states.

Nonlinear Causality

Nonlinear causality describes cause-effect relationships where outputs are disproportionate to inputs, involving feedback loops and potential abrupt shifts.

Observer Effect (Quantum)

The principle that the act of measurement influences the system being observed; metaphorically, the impact of metrics and monitoring on policy and behavior.

Path Dependency

The tendency for historical choices and system structures to shape and constrain future options.

Permafrost Thaw

The melting of long-frozen ground in polar regions, releasing stored carbon and methane

Phase Shift / Transition

A sudden, rapid, qualitative change from one state to another – often large-scale in system triggered by gradual pressure or small perturbations when a critical threshold is crossed.

Planetary Boundaries

Scientifically defined thresholds for critical Earth-system processes, beyond which abrupt or irreversible environmental change may occur.

Polycentric Governance

A governance model with multiple centers of decision-making authority operating autonomously but interacting and cooperating.

Positive Feedback Loop

A cycle that amplifies an initial change, often leading to runaway effects.

Precautionary Approach

A principle in environmental policy advocating preventive action in the face of uncertainty to avoid crossing dangerous thresholds.

Probability Amplitude (Quantum)

A mathematical expression of the likelihood of a particle's state; metaphorically, the spectrum of possible futures in sustainability planning.

Quantum Democracy

A participatory governance vision that balances decentralized decision-making with global coherence.

Quantum Superposition

A principle where a quantum system exists in multiple possible states simultaneously until observed; metaphorically, the coexistence of multiple possible policy pathways.

Rebound Effect

When efficiency improvements lower costs or barriers, prompting increased use that offsets some or all of the expected benefits.

Reciprocity (Governance)

Two-way feedback between decision-makers and those affected, ensuring accountability and trust.

Regenerative Design

A design philosophy that aims not only to minimize harm but to restore and enhance ecological and social systems.

Resilience

The capacity of a system to absorb disturbances or shocks, adapt, and recover its core functions to continue functioning.

Self-Organization

The spontaneous formation of order or patterns in a system without centralized control.

Scenario Planning

A strategic method for exploring multiple plausible futures to prepare for uncertainty.

Single-Use Plastics Ban Trade-offs

The unintended environmental costs of replacing plastic with other materials, such as paper or bioplastics.

Status Quo Bias

The tendency to prefer existing conditions and resist change, even in the face of evidence favoring alternatives.

Superposition (Metaphor)

Holding competing priorities or models simultaneously until conditions clarify the optimal choice.

System Boundaries

The conceptual or physical limits that define what is included in a system analysis.

Systemic Risk

The possibility that failure in one part of a system will cause cascading failures throughout the whole system.

Systems Thinking

An approach that considers the interconnections, feedback loops, and dynamics of the whole system rather than focusing on isolated components.

Systemic Mismatch

When policy or technical solutions are designed with linear assumptions but applied to nonlinear, adaptive systems.

Threshold Effect

A sudden change in a system's behavior once a certain limit is crossed, often irreversible within human timeframes.

Tipping Point

A threshold where a small change causes a disproportionate and often irreversible shift in system behavior.

Tunneling (Quantum)

A phenomenon where particles pass through barriers they classically should not cross; metaphorically, innovation pathways that bypass institutional obstacles.

Volatility Driver

A factor—technological, political, or narrative—that significantly increases uncertainty and rapid change in systems or markets.

References

1 Levin, Simon A. "Ecosystems and the Biosphere as Complex Adaptive Systems." *Ecosystems* 1, no. 5 (1998): 431–436. https://link.springer.com/article/10.1007/s100219900037

2 Wahal, S. "Decarbonizing the Pulp and Paper Industry: The Disruptive Promise of Gene-Edited Low-Lignin Trees." *International Fibers Journal*, no. 5 (2025). https://www.fiberjournal.com/decarbonizing-pulp-and-paper-industry/

3 Wahal, S. *The eCarbon Card Blueprint: Digital Solutions for Carbon Accountability – India as a Model.* Armin Lear Press, 2025. https://www.amazon.com/eCarbon-CardTM-Blueprint-Solutions-Accountability-ebook/dp/B0FP5V24B2

4 Maslin, Mark. *Climate Change: A Very Short Introduction*. 4th ed. Oxford: Oxford University Press, 2021.

5 Levin, "Ecosystems and the Biosphere," 431-436.

6 Lenton, T. M. "Tipping points in the climate system." *Weather* 76, no. 10 (2021). https://doi.org/10.1002/wea.4058

7 Global Tipping Points. "Positive Tipping Points." Accessed March 27, 2026. https://global-tipping-points.org/positive-tipping-points/

8 Lenton, T. M. "Early Warning of Climate Tipping Points." *Nature Climate Change* 1, no. 2 (2011): 201–209. https://doi.org/10.1038/nclimate1143

9 Lenton, T. M., et al., "Tipping Elements in the Earth's Climate System." *Proceedings of the National Academy of Sciences* 105, no. 6 (2008): 1786–1793. https://doi.org/10.1073/pnas.0705414105

10 Ditlevsen, P. D., and S. J. Johnsen. "Tipping Points: Early Warning and Wishful Thinking." *Geophysical Research Letters* 37, no. 19 (2010): L19703. https://doi.org/10.1029/2010GL044486

11 Ripple, William J., Christopher Wolf, Johan Rockström, Katherine Richardson, Nico Wunderling, Jay W. Gregg, Thomas Westerhold, and Hans Joachim Schellnhuber. "The Risk of a Hothouse Earth Trajectory." *One Earth* (2026). https://doi.org/10.1016/j.oneear.2025.101565.

12 Vuong, Q., V. P. La, and M. H. Nguyen. "Serendipity and Inherent Non-Linear Thinking Can Help Address the Climate and Environmental Conundrums." 2024. https://doi.org/10.32388/ULP2R8
13 Humboldt, Alexander von. *Essai sur la géographie des plantes (Essay on the Geography of Plants)*. Paris: Levrault, Schoell et Cie, 1805.
14 Humboldt, *Essai sur la géographie des plantes.*
15 Wulf, Andrea. *The Invention of Nature: Alexander von Humboldt's New World.* New York: Knopf, 2015.
16 Meadows, Donella H. *Leverage Points: Places to Intervene in a System.* Hartland, VT: Sustainability Institute, 1999.
17 Wahal, "Decarbonizing the Pulp and Paper Industry."
18 SeaCell™LT. *The Power of Seaweed in a Fiber.* Accessed February 13, 2026. https://algaefoundationatec.org/aces/library/The%20Power%20of%20Seaweed%20in%20a%20Fiber.pdf
19 Camilleri, E., S. Narayan, D. Lingam, and R. Blundell. "Mycelium-Based Composites: An Updated Comprehensive Review." *Composites Part C: Open Access.* Advance online publication, 2025. https://doi.org/10.1016/j.biotechadv.2025.108517
20 Wahal, S. "Mycelium-Based Biomaterials: Robust Material Platform and Sustainable Alternatives to Fossil and Biopolymer Foams." *International Fibers Journal*, no. 6 (2025). https://www.fiberjournal.com/mycelium-based-biomaterials/
21 Santos, M., M. Mariz, I. Tiago, S. Alarico, and P. Ferreira. "Bio-Based Polyurethane Foams: Feedstocks, Synthesis, and Applications." *Biomolecules* 15, no. 5 (2025): 680. https://doi.org/10.3390/biom15050680
22 Roychand, R., S. Kilmartin-Lynch, M. Saberian, J. Li, G. Zhang, and C. Q. Li. "Transforming Spent Coffee Grounds into a Valuable Resource for the Enhancement of Concrete Strength." *Journal of Cleaner Production* 419 (2023): 138205. https://doi.org/10.1016/j.jclepro.2023.138205
23 Vinall, Frances. "Cream and Sugar with Your Concrete? Australians Add Coffee to Cement." *Washington Post*, September 28, 2023. https://www.washingtonpost.com/world/2023/09/28/australia-coffee-grounds-cement-sustainability/

24 Westervelt, Amy. "Big Oil Firms Touted Algae as Climate Solution. Now All Have Pulled Funding." *The Guardian*, March 17, 2023. https://www.theguardian.com/environment/2023/mar/17/big-oil-algae-biofuel-funding-cut-exxonmobil
25 Buckley, S. "Algae May Be a 'Brilliant' Solution for Capturing Carbon at Gigaton Scale." *Sustainable Brands*, 2023. https://sustainablebrands.com/read/chemistry-materials-packaging/algae-may-be-a-brilliant-solution-for-capturing-carbon-at-giga-ton-scale
26 Moreira, D., and J. C. Pires. "Atmospheric CO_2 Capture by Algae: Negative Carbon Dioxide Emission Path." *Bioresource Technology* 215 (2016): 371–379. https://doi.org/10.1016/j.biortech.2016.03.060
27 Onyeaka, H., T. Miri, K. Obileke, A. Hart, C. Anumudu, and Z. Al-Sharify. "Review—Minimizing Carbon Footprint via Microalgae as a Biological Capture." *Carbon Capture Science & Technology* 1 (December 2021): 100007. https://doi.org/10.1016/j.ccst.2021.100007
28 Adetunji, A., S. Gumbi, and M. Erasmus. "Harnessing the Potential of Microalgae in Sequestration of CO_2 Emissions: Removal Mechanisms, Optimization Strategies, and Bioenergy Production." *Journal of Hazardous Materials Advances* 18 (2025): 100722. https://doi.org/10.1016/j.hazadv.2025.100722
29 CBS News. "One Farmer's Seaweed Discovery Could Help Slow Methane Emissions — and Change the World." 2021. https://www.cbsnews.com/news/seaweed-methane-emissions-cows-gas-climate-change/.
30 Editorial. "Benefits of Seaweed." *Nature Plants* 9 (2023): 1. https://doi.org/10.1038/s41477-023-01348-6
31 Malkiel, Burton G. *A Random Walk Down Wall Street*. New York: W. W. Norton & Company, 1973.
32 Wahal, *The eCarbon Card Blueprint*.
33 Wahal, S., and Kota, P. "Forecasting a United States Carbon Index Using Neural Networks and Economic Indicators." Paper presented at the AI Plus Symposium, University at Albany, New York, March 6, 2026.
34 Wahal, "Decarbonizing the Pulp and Paper Industry."

35 Wulf, The Invention of Nature.
36 Humboldt, *Essai sur la géographie des plantes.*
37 Maslin, *Climate Change: A Very Short Introduction.*
38 Ripple et al., "The Risk of a Hothouse Earth Trajectory."
39 Mulgan, Geoff. *Big Mind.* London: Princeton University Press, 2017. https://www.geoffmulgan.com/books/big-mind
40 Intergovernmental Panel on Climate Change (IPCC). *Climate Change 2013: The Physical Science Basis. Contribution of Working Group I to the Fifth Assessment Report of the Intergovernmental Panel on Climate Change.* Cambridge: Cambridge University Press, 2013. https://www.ipcc.ch/report/ar5/wg1/
41 "Braess's Paradox." *Wikipedia.* Accessed February 13, 2026. https://en.wikipedia.org/wiki/Braess'_paradox
42 Rial, J. A., R. A. Pielke, M. Beniston, M. Claussen, J. Canadell, P. Cox, et al. "Nonlinearities, Feedbacks and Critical Thresholds within the Earth's Climate System." *Climatic Change* 65, nos. 1–2 (2004): 11–38. https://doi.org/10.1023/B:CLIM.0000037493.89489.3f
43 Carbon Brief. "How Feedback Loops and Nonlinear Thinking Can Inform Climate Policy." 2023. https://www.carbonbrief.org/guest-post-how-feedback-loops-and-non-linear-thinking-can-inform-climate-policy/
44 Maslin, *Climate Change: A Very Short Introduction.*
45 Ripple et al., "The Risk of a Hothouse Earth Trajectory."
46 Rial et. al., "Nonlinearities, Feedbacks and Critical Thresholds," 11-38.
47 Rial et. al., "Nonlinearities, Feedbacks and Critical Thresholds," 11-38.
48 May, R. M. "Thresholds and Breakpoints in Ecosystems with a Multiplicity of Stable States." *Nature* 269, no. 5628 (1977): 471–477. https://www.nature.com/articles/269471a0
49 Maslin, *Climate Change: A Very Short Introduction.*
50 Ripple et al., "The Risk of a Hothouse Earth Trajectory."
51 "Emergence." *Wikipedia.* Accessed February 13, 2026. https://en.wikipedia.org/wiki/Emergence.
52 Asher, Claire, and Chris Forman. *Brave Green World: How Science Can Save Our Planet.* Cambridge, MA: MIT Press, 2021.
53 Yang, Y., K. Foster, K. Z. Coyte et al. "Time Delays Modulate

the Stability of Complex Ecosystems." *Nature Ecology & Evolution* 7 (2023): 1610–1619. https://doi.org/10.1038/s41559-023-02158-x
54 Biber, Eric. "Climate Change, Causation, and Delayed Harm." *Hofstra Law Review* 37, no. 4 (2009): 975–1005. https://law.hofstra.edu/pdf/academics/journals/lawreview/lrv_issues_v37n04_cc2_biber_final.pdf
55 Mulhern, Owen. "The Time Lag of Climate Change." *Earth.Org*, August 27, 2020. https://earth.org/data_visualization/the-time-lag-of-climate-change/
56 United States Environmental Protection Agency. "EPA Report Shows Disproportionate Impacts of Climate Change on Socially Vulnerable Populations in the United States." *EPA News Releases*, September 2, 2021. https://www.epa.gov/newsreleases/epa-report-shows-disproportionate-impacts-climate-change-socially-vulnerable.
57 Meyssignac, B., J. Chenal, N. Loeb, R. Guillaume-Castel, A. Ribes. "Time-Variations of the Climate Feedback Parameter λ Are Associated with the Pacific Decadal Oscillation." *Communications Earth & Environment* 4 (2023): Article 241. https://doi.org/10.1038/s43247-023-00887-2
58 "Heat Equation." *Wikipedia.* Accessed February 13, 2026. https://en.wikipedia.org/wiki/Heat_equation
59 Walker, Brian, and David Salt. *Resilience Practice: Building Capacity to Absorb Disturbance and Maintain Function.* Washington, DC: Island Press, 2012. https://islandpress.org/books/resilience-practice#desc
60 Walker and Salt, *Resilience Practice.*
61 Folke, Carl, Steve Carpenter, Brian Walker, Marten Scheffer, Terry Elmqvist, Lance Gunderson, and C. S. Holling. "Regime Shifts, Resilience, and Biodiversity in Ecosystem Management." *Annual Review of Ecology, Evolution, and Systematics* 35 (2004): 557–581. https://doi.org/10.1146/annurev.ecolsys.35.021103.105711.
62 Meadows, *Leverage Points.*
63 Folke et al., "Regime Shifts, Resilience, and Biodiversity," 557-581.
64 "Braess's Paradox," *Wikipedia.* .

65 Ditlevsen and Johnsen, "Tipping Points."
66 Meadows, *Leverage Points.*
67 Meadows, *Leverage Points.*
68 Levin, "Ecosystems and the Biosphere," 431–436.
69 Morrison, T.H., Adger, W.N., Agrawal, A. et al. Radical interventions for climate-impacted systems. *Nat. Clim. Chang.* 12, 1100–1106 (2022). https://doi.org/10.1038/s41558-022-01542-y.
70 Wahal, "Decarbonizing the Pulp and Paper Industry."
71 Morrison et al., "Radical Interventions."
72 Meadows, *Leverage Points.*
73 Wahal, *The eCarbon Card Blueprint.*
74 Anastas, Paul T., and Julie B. Zimmerman. "Creating Cascading Non-Linear Solutions for the UN Sustainable Development Goals through Green Chemistry." *Chem* 7, no. 10 (2021): 2825–2828. https://doi.org/10.1016/j.chempr.2021.10.025
75 Wahal, *The eCarbon Card Blueprint.*
76 Wahal, "Decarbonizing the Pulp and Paper Industry."
77 Camilleri et al., "Mycelium-Based composites."
78 Wahal, "Mycelium-Based Biomaterials."
79 Camilleri et al., "Mycelium-Based composites."
80 Wahal, "Mycelium-Based Biomaterials."
81 Wahal, *The eCarbon Card Blueprint.*
82 Kimmerer, Robin Wall. *Braiding Sweetgrass: Indigenous Wisdom, Scientific Knowledge, and the Teachings of Plants.* Minneapolis, MN: Milkweed Editions, 2013.
83 Wahal, *The eCarbon Card Blueprint.*
84 Wahal, "Decarbonizing the Pulp and Paper Industry."
85 Camilleri et al., "Mycelium-Based composites."
86 Wahal, "Mycelium-Based Biomaterials."
87 Wahal, "Decarbonizing the Pulp and Paper Industry."
88 Camilleri et al., "Mycelium-Based composites."
89 Wahal, "Mycelium-Based Biomaterials."
90 Eslacker, E., S. Vandelook, and E. Peeters. "Recent Technological Innovations in Mycelium Materials as Leather Substitutes: A Patent Review." *Frontiers in Bioengineering and Biotechnology* 11 (2023): 1204861. https://doi.org/10.3389/fbioe.2023.1204861
91 Amobonye, Ayodeji, J. Lalung, M. K. Awasthi, and S. Pillai.

"Fungal Mycelium as Leather Alternative: A Sustainable Biogenic Material for the Fashion Industry." *Sustainable Materials and Technologies* 38 (2023): e00724. https://doi.org/10.1016/j.susmat.2023.e00724

92 Crawford, A., S. R. Miller, S. Branco, J. Fletcher, and D. Stefanov. "Growing Mycelium Leather: A Paste Substrate Approach with Post-Treatments." *Research Directions: Biotechnology Design* 2 (2024): e6. https://doi.org/10.1017/btd.2024.6

93 Desserto. "Desserto – Technically Advanced Plant-Based Biomaterials." Accessed February 13, 2026. https://desserto.com.mx/.

94 Ananas Anam. "About Us." Accessed September 17, 2025. https://www.ananas-anam.com/about-us/.

95 Equinetree. "The Production of Uppeal™ Apple Leather." Accessed September 17, 2025. https://equinetree.com/magazine/what-is-uppeal-10-advantages

96 Samir, A., F. H. Ashour, A. A. Abdel Hakim, and M. Bassyouni. "Recent Advances in Biodegradable Polymers for Sustainable Applications." *npj Materials Degradation* 6 (2022): 68. https://doi.org/10.1038/s41529-022-00277-7

97 SeaCell™LT, *The Power of Seaweed in a Fiber*.

98 WellFabric. "Milk Fabric: Properties, Sustainability, and Applications." Accessed February 13, 2026. https://wellfabric.com/milk-fabric/

99 Gaminian, H., B. Ahvazi, J. J. Vidmar, U. Ekuere, and S. Regan. "Revolutionizing Sustainable Nonwoven Fabrics: The Potential Use of Agricultural Waste and Natural Fibres for Nonwoven Fabric." *Biomass* 4, no. 2 (2024): 363–401. https://doi.org/10.3390/biomass4020018

100 Thooshan. "Thooshan – Best Alternative to Single-Use Plastic and Compostable Tableware." Accessed February 13, 2026. https://www.thooshan.com/

101 Bombelli, P., C. J. Howe, and F. Bertocchini. "Polyethylene Bio-Degradation by Caterpillars of the Wax Moth *Galleria mellonella*." *Current Biology* 27, no. 8 (2017): R292–R293. https://doi.org/10.1016/j.cub.2017.02.060.

102 Sanluis-Verdes, A., P. Colomer-Vidal, F. Rodriguez-Ventura, M. Bello-Villarino, M. Spinola-Amilibia, E. Ruiz-Lopez, et al. "Wax Worm Saliva and the Enzymes Therein Are the Key to

Polyethylene Degradation by *Galleria mellonella*." *Nature Communications* 13, no. 1 (2022): 1–11. https://doi.org/10.1038/s41467-022-33127-w

103 Yoshida, S., K. Hiraga, T. Takehana, I. Taniguchi, H. Yamaji, Y. Maeda, et al. "A Bacterium That Degrades and Assimilates Poly(ethylene Terephthalate)." *Science* 351, no. 6278 (2016): 1196–1199. https://doi.org/10.1126/science.aad6359.

104 Ong, S. "The Living Things That Feast on Plastic." *ASBMB Today*, 2023. https://www.asbmb.org/asbmb-today/science/091023/the-living-things-that-feast-on-plastic

105 The Learning Network. "Nature's Solution to Plastic Pollution: The Amazing Power of the Wax Worm." *New York Times*, April 12, 2023. https://www.nytimes.com/2023/04/12/learning/natures-solution-to-plastic-pollution-the-amazing-power-of-the-wax-worm.html.

106 Santos et al., "Bio-Based Polyurethane Foams," 680.

107 Vinall, "Cream and Sugar with Your Concrete?"

108 Roychand et al., "Transforming Spent Coffee Grounds," 138205.

109 Roychand et al., "Transforming Spent Coffee Grounds," 138205.

110 Westervelt, "Big Oil Firms Touted Algae as Climate Solution."

111 Buckley, "Algae May Be a 'Brilliant' Solution."

112 Moreira and Pires, "Atmospheric CO_2 Capture by Algae," 371-379

113 Onyeaka et al., "Minimizing Carbon Footprint via Microalgae," 100007.

114 Adetunji, Gumbi, and Erasmus, "Harnessing the Potential of Microalgae," 100722.

115 CBS News, "One Farmer's Seaweed Discovery."

116 "Benefits of Seaweed," .

117 Westervelt, "Big Oil Firms Touted Algae as Climate Solution."

118 Westervelt, "Big Oil Firms Touted Algae as Climate Solution."

119 Buckley, "Algae May Be a 'Brilliant' Solution."

120 Moreira and Pires, "Atmospheric CO_2 Capture by Algae," 371-379

121 Onyeaka et al., "Review - Minimizing Carbon Footprint," 100007.

122 Mastercard. "DO Black: The World's First Credit Card with

a Carbon Limit." April 2019. https://www.mastercard.com/news/europe/sv-se/nyhetsrum/pressmeddelanden/sv-se/2019/april/do-black-the-world-s-first-credit-card-with-a-carbon-limit/.

123 Thales Group. "Inspired by Doconomy." Accessed September 17, 2025. https://www.thalesgroup.com/en/markets/digital-identity-and-security/banking-payment/inspired/doconomy.

124 *Times of India*. "West Bengal to Introduce World's First Carbon Credit Cards." January 2025. https://timesofindia.indiatimes.com/city/kolkata/west-bengal-to-introduce-worlds-first-carbon-credit-cards/articleshow/117318806.cms.

125 Sigma Earth. "Bengal Government to Launch World's First Carbon Credit Card for Individuals." January 2025. https://sigmaearth.com/bengal-government-to-launch-worlds-first-carbon-credit-card-for-individuals/.

126 Wahal and Kota, "Forecasting a United States Carbon Index."

127 California Air Resources Board. "Cap-and-Trade Program." Accessed February 13, 2026. https://ww2.arb.ca.gov/our-work/programs/cap-and-trade-program.

128 Regional Greenhouse Gas Initiative. "Program Overview and Design: Elements." Accessed February 13, 2026. https://www.rggi.org/program-overview-and-design/elements.

129 Washington State Department of Ecology. "Climate Commitment Act." Accessed February 13, 2026. https://ecology.wa.gov/air-climate/climate-commitment-act.

130 European Commission. "EU Emissions Trading System (EU ETS)." Accessed February 13, 2026. https://climate.ec.europa.eu/eu-action/carbon-markets/eu-emissions-trading-system-eu-ets_en.

131 California Air Resources Board, "Cap-and-Trade Program."

132 Regional Greenhouse Gas Initiative, "Program Overview and Design."

133 Wahal and Kota, "Forecasting a United States Carbon Index."

134 Lorenz, Edward N. "Deterministic Nonperiodic Flow." *Journal of the Atmospheric Sciences* 20, no. 2 (1963): 130–141. https://doi.org/10.1175/1520-0469(1963)020<0130:DNF>2.0.CO;2

135 Petersdorff, T. "Lorenz Attractor and Chaotic Systems." University of Maryland GRACE Center. Accessed February 13, 2026. https://www.grace.umd.edu/~petersd/lorenz.html.

136 Lamont-Doherty Earth Observatory. "Nonlinear Climate Variability and Predictability." Columbia University. Accessed February 13, 2026. https://ocp.ldeo.columbia.edu/res/div/ocp/arch/nonlinear.shtml.

137 Collett, M., J. Rosenow, S. Sharpe. "Guest Post: How Feedback Loops and Non-Linear Thinking Can Inform Climate Policy." *Carbon Brief*, September 7, 2020. https://www.carbonbrief.org/guest-post-how-feedback-loops-and-non-linear-thinking-can-inform-climate-policy/.

Additional Resources

Further Reading & Web Resources

The quantum journey toward sustainability demands both intellectual inquiry and practical experimentation. While this book provides a conceptual and strategic framework, the resources below offer deeper dives into specific domains—including systems thinking, climate dynamics, quantum metaphors, digital tools, and bio-based innovations.

These books, articles, platforms, and networks will support readers who wish to translate insight into action, explore emergent paradigms, or design interventions in their own spheres of influence.

Systems Thinking, Nonlinearity, and Complexity

- Meadows, D. H. (2008). *Thinking in Systems: A Primer.* Chelsea Green Publishing.
  (A timeless introduction to systems thinking and leverage points.)
- Capra, F., & Luisi, P. L. (2014). *The Systems View of Life: A Unifying Vision.* Cambridge University Press.
  (Explores how living systems theory can reshape our approach to ecology, economy, and culture.)
- Holland, J. H. (2014). *Complexity: A Very Short Introduction.* Oxford University Press.
  (A concise overview of complexity science and its relevance to adaptive systems.)
- Arthur, W. B. (2014). *Complexity and the Economy.* Oxford University Press.
  (A deep dive into how technologies evolve through combinatorial processes rather than linear invention.)
- Taleb, N. N. (2012). *Antifragile: Things That Gain from Disorder.* Random House.

(A framework for thriving amid uncertainty, applicable to both systems and organizations.)

- Ostrom, E. (1990). *Governing the Commons*. Cambridge University Press.
  (Nobel Prize-winning analysis of cooperative resource management and emergent governance models.)
- West, G. (2017). *Scale: The Universal Laws of Growth, Innovation, Sustainability, and the Pace of Life*. Penguin Press.
  (Insights into the mathematical patterns underlying cities, companies, and ecosystems.)
- Number Analytics: Navigating Nonlinear Environmental Ethics
  (A resource for applying quantitative tools to analyze ethical and sustainability trade-offs in nonlinear systems.)
- Sustainability Directory: Nonlinear Causality and Sustainability Dynamics
  (A curated index of research, tools, and organizations linking causality science to sustainability practice.)

Quantum Thinking and Metaphorical Extensions

- Bohm, D. (1980). *Wholeness and the Implicate Order*. Routledge.
  (A classic text on interconnectedness and holistic thinking.)
- Barad, K. (2007). *Meeting the Universe Halfway*. Duke University Press.
  (Explores how quantum entanglement can inform ethics, epistemology, and sustainability.)
- Gribbin, J. (1984). *In Search of Schrödinger's Cat*. Bantam.
  (An accessible explanation of quantum theory, its paradoxes, and implications.)
- Rovelli, C. (2021). *Helgoland: Making Sense of the Quantum Revolution*. Riverhead Books.
  (A philosophical exploration of quantum mechanics as relational rather than deterministic.)
- Wheatley, M. J. (2006). *Leadership and the New Science*. Berrett-Koehler.

(Applies principles from quantum physics, chaos theory, and complexity to leadership and organizational change.)

- Baggott, J. (2011). *The Quantum Story: A History in 40 Moments.* Oxford University Press.
  (A narrative journey through the key breakthroughs that shaped quantum science.)
- Vuorio, J. (2024). *Serendipity and Nonlinear Thinking in the Climate Crisis.*
  (Examines how chance discoveries and nonlinear problem-solving can drive climate innovation.)

Climate Crisis and Sustainability Frameworks

- IPCC. (2021–2023). *Sixth Assessment Reports* (AR6). Intergovernmental Panel on Climate Change.
  (Comprehensive scientific consensus on climate dynamics, mitigation, and adaptation pathways.)
- Rockström, J. et al. (2009). *A Safe Operating Space for Humanity. Nature*, 461, 472–475.
  (Foundational work defining planetary boundaries for sustainable human activity.)
- Raworth, K. (2017). *Doughnut Economics.* Chelsea Green Publishing.
  (A transformative economic framework balancing ecological ceilings and social foundations.)
- Hawken, P. (Ed.). (2017). *Drawdown.* Penguin Books.
  (A ranked compendium of the most effective solutions to reverse global warming.)
- Hawken, P. (2021). *Regeneration.* Penguin Books.
  (A blueprint for holistic climate action focused on healing ecosystems and communities.)
- Ghosh, A. (2016). *The Great Derangement.* University of Chicago Press.
  (A critique of cultural and political blind spots in addressing the climate crisis.)

- Figueres, C., & Rivett-Carnac, T. (2020). *The Future We Choose*. Vintage.
  (A practical guide to making decisive choices that shape a livable climate future.)

Behavioral Science, Leadership, and Governance

- Kahneman, D. (2011). *Thinking, Fast and Slow*. Farrar, Straus and Giroux.
  (A landmark exploration of the two systems that drive human decision-making.)
- Thaler, R. H., & Sunstein, C. R. (2009). *Nudge*. Penguin Books.
  (How small design changes in choice architecture can lead to significant behavioral shifts.)
- Senge, P. (2006). *The Fifth Discipline*. Doubleday.
  (A guide to building learning organizations through systems thinking.)
- Scharmer, O., & Kaufer, K. (2013). *Leading from the Emerging Future*. Berrett-Koehler.
  (Introduces "Theory U" as a framework for transformational leadership in complex systems.)
- Donella Meadows Institute: www.donellameadows.org
  (A hub for resources, case studies, and training on systems thinking and sustainability.)
- Mindsponge: AI and Nonlinear Thinking in Climate Science
  (An emerging platform using AI-driven cognitive models to analyze climate complexity.)

Bio-Based Innovation and Nature-Based Solutions

- Wahal, S. (2025). *Decarbonizing the Pulp and Paper Industry*. Decarbonization LLC.
  (Explores the disruptive potential of gene-edited low-lignin trees to cut emissions and improve efficiency.)
- FAO Reports on Agroecology and Regenerative Agriculture

(Global policy and practice guides for climate-resilient, biodiversity-friendly agriculture.)

- Mycelium-Based Materials: Ecovative Design, MycoWorks
  (Case studies of companies scaling nature-grown packaging, textiles, and composites.)
- CRISPR in Lignin Engineering: *Molecules and Nature Biotechnology*
  (Key research articles on gene-editing pathways for industrial decarbonization.)
- Ellen MacArthur Foundation: www.ellenmacarthurfoundation.org
  (Thought leadership and tools for accelerating the transition to a circular economy.)
- International Fiber Journal and Sustainable Nonwovens

(Industry publications tracking innovations in fiber, nonwoven, and sustainable textile technologies.)

Digital Tools, Metrics, and Climate Simulators

- Our World in Data: www.ourworldindata.org
- WRI Climate Data Explorer: www.climatewatchdata.org
- Carbon Pricing Dashboard (World Bank): carbonpricingdashboard.worldbank.org
- Climate Interactive: www.climateinteractive.org
- MIT Climate Pathways Project
- OpenLCA and GHG Protocol Tools for life-cycle and emissions accounting

Global Platforms and Practitioner Resources

- Project Drawdown: www.drawdown.org
- UNEP Emissions Gap Reports: www.unep.org/resources/emissions-gap-report
- Circularity Gap Report: www.circularity-gap.world
- Climate Action Tracker: www.climateactiontracker.org

- IETA – International Emissions Trading Association: www.ieta.org

Networks and Communities of Practice

- ICLEI – Local Governments for Sustainability: www.iclei.org
- C40 Cities: www.c40.org
- Global Covenant of Mayors: www.globalcovenantofmayors.org
- B Corp Movement: www.bcorporation.net
- Regeneration.org – Paul Hawken's community-based climate solutions

Future-Oriented Thinking and Design Futures

- Deutsch, D. (2011). *The Beginning of Infinity*. Penguin.
- Kelly, K. (2010). *What Technology Wants*. Viking.
- Rifkin, J. (2019). *The Green New Deal*. St. Martin's Press.
- Wheatley, M. (2006). *Leadership and the New Science*. Berrett-Koehler.

Online Journals and Research Gateways

- *Nature Climate Change*
- *Climatic Change*
- *Environmental Research Letters*
- *Energy Policy*
- *Journal of Industrial Ecology*
- *Carbon Management Journal*

Exercises & Tools from This Book

To extend the book's frameworks into practice:

- Quantum Walk Canvas: Map entangled strategies, feedback loops, and uncertainties for decision-making.

- Entanglement Maps: Visualize cross-sectoral effects of technologies or policies.
- Scenario Constellations: Explore multiple plausible futures rather than fixed forecasts.
- Feedback Dashboards: Build real-time monitoring systems for social, ecological, and emotional signals—not just numbers.

Acknowledgments

Writing this book has been both an intellectual journey and a deeply personal one. It would not have been possible without the many people and organizations who have supported, challenged, and inspired me along the way.

First, I wish to express my deepest gratitude to my family—my wife, Anupama, and our daughters, Shweta and Arpita—whose love, patience, and unwavering belief in me made this work possible. You have been my anchor and my compass, gently reminding me why this work matters and encouraging me through every late night and early morning.

I am indebted to the many colleagues, mentors, and collaborators I have had the privilege to work with over the years. Your insights and diverse perspectives have shaped my thinking, sharpened my arguments, and expanded the horizons of what I believed possible.

Special thanks go to my friends and peers in the sustainability, innovation, and leadership communities. Your openness to share ideas, debate frameworks, and test unconventional approaches has enriched every chapter of this book. The cross-pollination of science, policy, and lived experience is the essence of systems thinking, and you have all contributed to that richness in ways too numerous to count.

I also wish to acknowledge the countless researchers, writers, and practitioners whose work I have drawn upon. Whether through books, articles, conversations, or conference panels, your contributions have provided the intellectual scaffolding upon which I could build.

To the publishing and editorial teams of Cal Poly Humboldt University—thank you for your guidance, precision, and patience. Your ability to see both the forest and the trees has ensured that the book is as clear, compelling, and useful as I hoped it would be.

Finally, I want to thank every reader—whether you are a policy-maker, an entrepreneur, a student, or simply someone curious about reimagining our future. This book is written for you. If it sparks a new insight, prompts an important conversation, or inspires you to act in ways you had not considered before, then it will have fulfilled its purpose.

For all those named and unnamed, near and far, thank you for walking this journey with me.

— Sanjay Wahal, PhD

About the Author

Turning science, systems thinking, and human behavior into climate-positive transformation.

Dr. Sanjay Wahal is a sustainability strategist, systems thinker, and climate innovator working at the intersection of decarbonization technologies, materials innovation, and circular economy solutions. With a PhD in Chemical Engineering, an MBA in Strategy and Innovation, and an undergraduate degree from IIT Kanpur, he brings over three decades of cross-functional leadership spanning industrial R&D, advanced materials, clean energy, sustainability strategy, and climate-tech innovation—bringing rare clarity to complex global challenges.

Dr. Wahal is the Founder and President of Decarbonization, LLC, an advisory firm dedicated to helping companies, investors, and institutions accelerate the transition to low-carbon and circular economic systems. Over the course of his career, he has advised global corporations, startups, and institutional investors on climate innovation, clean technologies, circular economy models, the transition toward regenerative systems, and commercialization pathways for sustainable materials and industrial technologies.

He is the author of *The eCarbon Card Blueprint: Digital Solutions for Carbon Accountability – India as a Model* (Armin Lear Press, 2025), along with several forthcoming books exploring profitable circular economy innovations, nonlinear leadership, quantum mindsets, planetary-boundaries constraints, and the intersection of culture, leadership, and climate transformation.

Dr. Wahal has also proposed innovative frameworks such as the United States Carbon Allowance (USCA) Index, which links macroeconomic indicators to carbon price forecasting. He continues to collaborate with climate-tech ventures and impact investors to accelerate decarbonization through innovation, finance, and systems design.

Beyond his technical and industrial contributions, his writing explores the profound convergence of ancient wisdom and modern science in everyday human practices—how we eat, walk, work, and lead. Drawing from complexity science, behavioral economics, yoga philosophy, and cultural traditions, his work invites readers to reimagine sustainability not merely as a technological or policy challenge, but as a deeply human endeavor rooted in behavior, consciousness, and lived experience.

Learn more at:
www.decarbonizationllc.com
www.sanjaywahal.com